Jw.

P/9.95
D3

D0216151

INVITATION
TO PHYSICS

Ken Greider
University of California, Davis

INVITATION TO PHYSICS

Harcourt Brace Jovanovich, Inc.

New York Chicago San Francisco Atlanta

Drawings by Ed Malsberg
Figures 1-1, 2-1, 4-8, 5-7, 6-6, 6-7, 6-9, 7-3, 7-4, 7-6, 14-1, 14-2, 14-3, 14-4, 15-1, 15-2, 15-3, 15-5, 15-6, 15-7, 15-8, 15-9, 15-10, 16-3, 16-4, 16-6.

© 1973 by Harcourt Brace Jovanovich, Inc.

All rights reserved. No part of this publication may be reproduced or transmitted in any form or by any means, electronic or mechanical, including photocopy, recording, or any information storage and retrieval system, without permission in writing from the publisher.

ISBN: 0-15-546907-X

Library of Congress Catalog Card Number: 72-917-46

Printed in the United States of America

To my Mother and my Father

Preface

Invitation to Physics is intended for a one-semester or one-quarter course in elementary physics for humanities students, or for the physics segment of a general introduction to physical science. It is the final product of a series of lecture notes that I developed and class tested for several years at the University of California at Davis.

I had two objectives in writing this book: (1) to expose the reader to the changing nature of science in general through a study of the evolution of physics in particular and (2) to recreate the changing ideas of physics as real revolutions in man's perception of the universe—revolutions that were (and still are) being made by real men.

Although it has been my experience in teaching this course that some degree of problem solving is beneficial, I have avoided any temptation to emphasize mathematical manipulations. A blend of mathematical problems and/or questions requiring nonmathematical answers appears at the end of each chapter for use at the instructor's discretion. The few mathematical exercises that have been included are designed to enhance the student's understanding of the subject and allow the reader to recreate a degree of the emotional satisfaction that a physicist feels when he successfully brings theory to bear on solving a practical problem. If I have succeeded in communicating any of the excitement, the creativity, and the fun of stretching one's mind with new ideas, the humanities student may come to appreciate physics as a science that is both comprehensible and enlightening.

In addition to mathematical and verbal problems, the end-of-chapter material includes a few questions requiring outside reading. The instructor can use these questions as he wishes for term-paper subjects or as topics for small-group discussions.

I have tried to free the main body of the book from lengthy mathematical derivations that would detract from its readability. Several appendixes contain these derivations for the mathematically curious student.

I have included a number of references to and diagrams of lecture demonstrations that are common in any beginning physics course. Most of these are available commercially or are easy to build, and I strongly recommend their use as supplementary visual aids. The first appearance of new units, technical definitions, and terms is rendered in color merely to emphasize the fact that they are likely to be new to the reader. Brief biographical portraits of 25 key physicists are included.

In developing this book, I was indebted to the work of many hands and the inspiration of many minds. I would especially like to express my appreciation and gratitude to: Dave Faulkner for his suggestions and for the countless, tedious hours he spent helping to prepare questions and problems. Frank Serduke for his assistance on earlier versions of the manuscript. Virginia Rosato and Rory Tafoya, who typed the notes and the manuscript, for their patience in decoding my scribbling. My reviewers who read the manuscript in its earlier stages—William J. Bruff, Foothill College, Los Altos Hills, California; John D. McCullen, The University of Arizona; and Paul W. Wagner, Los Angeles City College—for contributing helpful comments and suggestions that have improved the book immensely. Hans Weidenmüller for the kind hospitality of the Max Planck Institut für Kernphysik in Heidelberg, West Germany, where the last stages of the book were completed. Roger Dunn for his prodding and patience. Marian Flanders for her warm support and understanding throughout all my writing agonies. And, last, my many, many friends who, over the years as my students, contributed immeasurably to the development of this book.

Ken Greider

Contents

1

2

Magnetism and Electromagnetic Induction 132

10

Light 144

11

Quantum Mechanics: A New Picture of the Universe 235

18

Nuclear Physics — An Application of Quantum Mechanics 256

19

Appendixes   297

Index   307

Early Greek Physics

From the beginning of recorded time, we know that man has wondered about his universe. Even the ancients speculated on the origins and causes of natural phenomena. Why *do* the planets, stars, sun, and moon move across the sky the way they do? Why *do* objects on the earth rise and fall? What *is* the "stuff" from which everything is made? Finding the answers to such questions is what physics is all about.

Many of the natural phenomena still overwhelming to us were especially frightening to the ancients. Lightning, thunder, eclipses, comets, and similar occurrences were given magical or divine origins. In the mythology of ancient times, gods like Thor, Logi, and Zeus were the causes of natural phenomena. By the sixth century B.C., however, a new awareness had become evident in ancient Greece, an awareness which was to lead to the development of a new philosophy in which man began to reflect—to sit back and look again at nature, at the universe—in a different way and with a certain detachment. Looking at the universe objectively for the first time, men began to construct theories regarding its composition and structure. Logic and mechanistic models soon replaced mysticism and magic as men sought to understand and explain their universe.

The ancients originally became interested in astronomy and in the celestial motion of heavenly bodies to meet their immediate practical needs (navigation, timekeeping, and so on). The early Babylonians and Egyptians formulated calendars based on observed astronomical cycles. It seems natural to suppose that once

1-1

Introduction

1-2

The Early Greek Period: Sixth and Fifth Centuries B.C.

the regularities of the stars and planets had been established, someone would produce a mechanistic model or a rational explanation that could predict these regularities. As early as the sixth century B.C., recorded speculations about such explanations existed—speculations which were to continue through the centuries and even to the present day.

Thales of Miletus was one of the first to propose the existence of a basic element from which all things in the universe were created and to look for principles or laws under which the world operated. Unfortunately, little first-hand evidence of the writings of Thales, as well as of the other early philosophers, was preserved; most of the earliest ideas about the universe come to us second- or third-hand from later philosophers such as Aristotle and commentators such as Simplicius. The early Milesian school, influenced primarily by Thales, Anaximander, and Anaximenes, among other things speculated about what "held up" the earth. Anaximenes thought the earth rested on air; Thales, on water. Such natural explanations for the support of the earth represented a substantial break with mythological causes and the premise that Atlas held up the earth on his back. Anaximander broke even more spectacularly with traditional thought when he postulated that the earth was a body in the center of the universe which remained stationary because it was equidistant from all other things. But how was such an earth "held up?" Clearly, a new revolutionary idea like Anaximander's raised further questions that had to be answered. A whole cosmology or mechanism of the universe was demanded. The motion of the sun, moon, planets, and stars now had to be explained without the use of mythological agents.

It is said that Anaximander also believed the earth to be drum-shaped, with one of the flat ends providing the surface we live on as shown in Figure 1-1. The first known description of a spherical universe comes from Parmenides of Elea (now Italy) in the early part of the fifth century B.C. who said the universe was "like the mass of a well-rounded sphere, from the middle equal in every respect." This picture of an earth-centered universe was the basis for later Greek (particularly Aristotelian) cosmology and remained a fundamental principle in both physics and astronomy until the end of the sixteenth century A.D.

Other early Greek philosophers speculated on the constitution of matter and on the "forces" that cause change. In the late sixth century B.C., Heraclitus of Ephesus thought the basic element in the universe was fire and that the force of change was due to strife between opposites (war and peace, day and night, and so on). This same theme was elaborated upon in the early fifth century B.C. by Empedocles, who postulated that Love and Strife were the two

1-1 Anaximander's drum-shaped world.

forces which caused changes in what he proposed were the four basic elements: fire, air, water, and earth. Love united and Strife separated these elements into various forms which man saw in everything in the world around him. These four elements were also the basis of Aristotelian teaching 100 years later and remained the basic "stuff" of the physical world for 2000 years.

It is interesting to note, however, that even during this early period there were competing ideas about the structure of matter as well as the structure of the universe. In particular, Anaxagoras, a contemporary of Empedocles and teacher of Pericles and Euripedes, thought there were an infinite number of elements or seeds, each of which contained matter from every material substance. In Anaxagoras' picture, Mind was the controlling force between the elements. Due to such revolutionary ideas, however, he was banished from Athens for teaching heresy! He was not the last scientist to be punished for unorthodox ideas.

Other philosophers, notably the atomists of the late fifth century, Leucippus and Democritus, held still another view. They proposed that all matter was composed of tiny invisible atoms which had varying shapes and hooked together to form the solids, liquids, and gases experienced in nature:

> By convention sweet is sweet, bitter is bitter, hot is hot, cold is cold, and color is color. But in reality there are atoms and the void. That is, the objects of sense are supposed to be real and it is customary to regard them as such, but in truth they are not. Only the atoms and the void are real.

Democritus believed the sensations and the qualities of objects (color, smoothness, sound) arose from the shape of the atoms and the particular way they hooked together and that, cosmologically, the universe was an infinite empty space, with the stars and the planets (composed of atoms, too) occupying positions in that space. According to him, the moon was very much like the earth, having mountains, valleys, plains, and seas. As we shall see, Democritus' view was abandoned in Aristotle's teaching and was not revived until 2000 years later when Galileo actually observed the mountains on the moon through a telescope.

1-3
The Pythagoreans

The Pythagoreans were also important in the development of cosmological ideas near the end of the fifth century B.C. The famous mathematician Pythagoras himself left no writings, and most of the information we have today is from his followers or disciples. His school was close to a religious order or commune where members adhered to a set of strict communal laws. According to them, numbers were the basic "stuff" of the universe—

PYTHAGORAS

Pythagoras holds the distinctions of being the inventor of the word "philosopher" and the developer of Greek communes. He is also famous for the Pythagorean Theorem which almost everyone learns in high school mathematics. □ Born in Samos somewhere between 609 B.C. and 570 B.C., Pythagoras spent a good deal of time traveling and learning from the major cultures of the age. Little is known about his life; reality is confused with myth on this point. His followers considered him half-god, a quality difficult to verify now. □ Legend tells of the first use of the word "philosopher." Pythagoras was at the Olympic games and a famous prince asked him what he did for a living. He explained that he was a philosopher. When asked what a philosopher was, Pythagoras answered that life was like the games: Some played because of the money, some because of the fame and glory; but there were some who only observed and tried to understand everything that went on. □ The communal society which he founded had strong religious overtones; rites included fasting and meditation exercises. Really a religious society, the Pythagorean commune was more concerned with moral reform than with philosophical teaching. Other communes grew from the first, and their influence on Greek politics where they were established was fairly powerful. The Pythagorean societies were heavily persecuted after they suggested, at the end of one war between two city-states, that the spoils of war should be public property. □ Pythagoras died at the beginning of the fifth century B.C., probably by choice at the end of a 40-day fast. According to legend, when the Roman Senators were later directed by an oracle to raise a monument to the wisest man who had ever lived, they built it for Pythagoras. The Pythagorean order continued to be powerful in Greece until the middle of the fifth century; it was then violently stamped out and the few surviving Pythagoreans were forced to flee the country.

Biblioteca Ambrosiana, Milan

Pythagoras; sketch by Raphael for "The School of Athens".

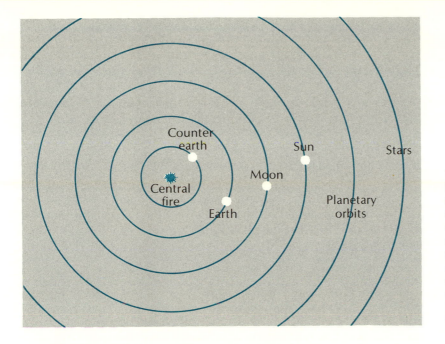

1-2 The Pythagorean picture of the universe. The inhabited part of the earth is always oriented outward so the central fire and counter earth are never observable. The earth does not rotate on its axis, but in its orbit around the central fire, thus explaining the rising and setting of the sun. The other celestial bodies also orbit the central fire, but at different rates.

numbers were divine and were everything. The mysticism of numbers played a dominant role in their ideas.

The Pythagoreans' cosmological model consisted of a spherical earth rotating in a circular motion—not about the sun, but about the central fire. In this model, the inhabited side of the earth always faced away from this fire, making it not visible to man. In addition, there was a "counter earth," also unobservable to man, rotating between the central fire and the earth. The sun and stars also rotated around the central fire, but their orbits were farther out. In this model (see Figure 1-2), the daily circular motion of the earth around the fire compared with the slower motion of the sun could explain the occurrence of night and day in a natural way.

Another novel feature of the Pythagorean cosmology was its connection with music. The Pythagoreans had already discovered that musical intervals were based on ratios of the whole numbers (an octave is 2:1; a major fifth is 3:2; a major third, 5:4; and so on). They extended this analogy to the solar system, proposing that the ratios of the various distances of the heavenly bodies from the central fire produced these same harmonies. The reason that these harmonies were not heard was the fact that man was not conscious of them, since they had existed from his birth. Unfortunately, this model for the distances between heavenly bodies did not agree with later, more accurate astronomical observations. The Pythagoreans' approach was to explain the regularities in nature by means of simple, natural, geometrical, and arithmetic relations. This method has been used throughout the history of astronomy and physics to help man organize and understand the world around him. Many times, as with the Pythagorean theory, the results did not fit future, more accurate experimental data. But often this method, which we will call "numerology," has been very successful in obtaining an empirical order in a set of confusing experimental results, as will be discussed in later chapters. Some scientists feel this approach is too mystical and should not be a part of normal science; others, however, who believe in a fundamental simplicity in the basic mathematical description of the universe, are willing to accept that much of nature can be described by such simple mathematical relations. The test, of course, is whether or not the theory based on numerology accurately fits observed experiments.

1-4
Plato

Around the time that the Pythagoreans were living in communes in Italy and Sicily, the school of Socrates was established in Athens. Important contributions to astronomy came from Plato (428–347 B.C.), who studied with Socrates and who urged astronomers to use quantitative mathematical descriptions of their work, instead of the qualitative pictures used previously. Plato also taught that astronomy, like arithmetic and grammar, was a subject that every schoolboy ought to learn.

Most of Plato's other ideas in physics came from previous pre-Socratics. Like Empedocles, he believed in the four elements of fire, earth, air, and water, but Plato added a fifth—the ether. Plato used many of the elements of earlier astronomers: the spherical universe of Parmenides, the circular orbits of the Pythagoreans, and the central earth of Anaximander. Although several members of Plato's Academy did postulate a rotating earth, there is no evidence that Plato himself believed in one.

Eudoxus (395–342 B.C.), a contemporary of Aristotle, was one of the first followers of Plato to comprehend and construct mathematical models of concentric celestial spheres with the earth at the center. Eudoxus' model (see Figure 1-3) accounted for complicated planetary motions, predicted eclipses, and was much more sophisticated mathematically than any previous concept. It served as a prototype for the rather successful model of an earth-centered (**geocentric**) universe developed by Aristotle and Hipparchus and refined in detail by Ptolemy several hundred years later.

Reviewing the main contributions of the pre-Aristotelians discussed here, (1) they originated most of the basic ideas on cosmology and physics that were to be developed in detail later and (2) they abandoned the traditions of religious dogma, mys-

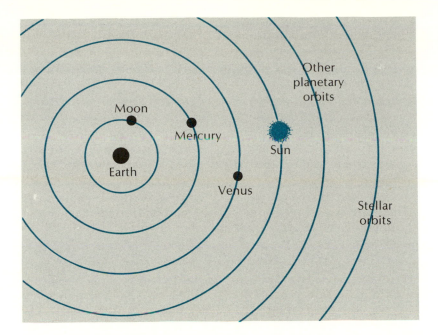

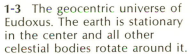

1-3 The geocentric universe of Eudoxus. The earth is stationary in the center and all other celestial bodies rotate around it.

ticism, and mythology to explain the world around us and started using logic and reason to give qualitative and descriptive pictures of natural phenomena. In pre-Aristotelian thought, the beginnings of the two main theories on the constitution of matter can be seen: Empedocles' proposal that everything in the universe was composed of one or more of the four elements of fire, air, water, and earth versus Democritus' theory that atoms were the basic constituents of matter and that everything else was an empty vacuum. The essences of these two models—the "plenum" of a space completely filled everywhere in the former case versus the particles in empty space in the latter—have been used over and over again in competing theories describing various physical phenomena, even to the present day.

The many subtle changes of thought and the difficulties in overcoming traditional dogma and mysticism posed enormous problems in the sixth to the fourth centuries B.C.—problems similar to the difficulties that the "new" physicists of the seventeenth century were to have in overthrowing the traditions of Aristotelian physics, as we will see in the next several chapters.

Questions

1 Describe briefly some of the ideas of the ancient Greeks concerning the physical nature of the world. Why do you think they had good or poor cause for believing as they did?

2 How would you expect "numerology" to lead to a valid description of a physical process?

3 Does the failure of the Pythagorean harmonies to describe the solar system mean that there are no regularities in the distances between planets? Explain.

4 What types of physical phenomena are commonly referred to today in terms of a plenum theory? In terms of particles (atoms) and empty space?

5 Why did thinkers beginning with Thales speculate on and propose the existence of some basic "stuff" (elements, atoms) from which all things in the universe are composed? Explain.

6 What do you think constitutes a valid theory? (That is, what attributes must a theory possess before we can accept it?) Explain.

7 What do you think the proper role of "common sense" should be in formulating physical theories and in supplying arguments, pro and con, about them? Is there really such a thing as "common sense?" Explain.

8 Answer Question 7 for "intuition."

9 What criteria do you think an observation must meet before you consider it a scientific fact? Reexamine your answers to Questions 6, 7, and 8 in light of your answer to this one. Have you changed your mind about any of your answers?

Questions Requiring Outside Reading

1 What kinds of actual experimental or observational evidence did the ancient Greeks employ? What actual observations did they make?

2 Find more detailed descriptions of the early atomists, especially Democritus and Leucippus. Do you think that these men had as good grounds for their beliefs as Plato and Aristotle did for their fire-earth-air-water scheme?

3 Find some of the mythological and mystical "causes" of natural phenomena believed in by early civilizations.

4 What was the purpose of the "counter earth" in the Pythagorean universe?

5 How did the Pythagorean model of the universe explain eclipses?

Aristotle's Physics   2

Aristotle (384–322 B.C.) was a student in Plato's Academy who advocated and developed the ideas of Socrates and Plato more than anyone else. As far as physics and astronomy were concerned, he had the greatest influence on the future of all the Greeks. Following Socrates and Plato, he subscribed to an organic view of the universe. Plato had said, "The world in very truth is a living creature with soul and reason." Aristotle added a similar view, "As in human operations, so in natural processes. Human operations are for an end; hence natural processes are so too." This view of a universe with lifelike purposes led Aristotle astray, particularly in the fields of physics and astronomy. Aristotle's errors would be only of academic interest now were it not for his enormous influence which affected the thinking of scientists and laymen alike for the next 2000 years. It took tremendous effort in both the sixteenth and seventeenth centuries to overthrow Aristotle's original view of the physical world. But before going into that, let us briefly review his main theories in physics and astronomy.

Introduction

Aristotle in particular started the ordered system of knowledge based on causes and effects that we use today, although other Greek philosophers were involved as well. He was able to generalize, categorize, and find a unity in the wide spectrum of distinct experience found in nature. In seeking this unity, however, Aristotle insisted that the various sciences, including physics and astronomy, were to be studied separately and, furthermore, that

2-2

Aristotle's Cosmology

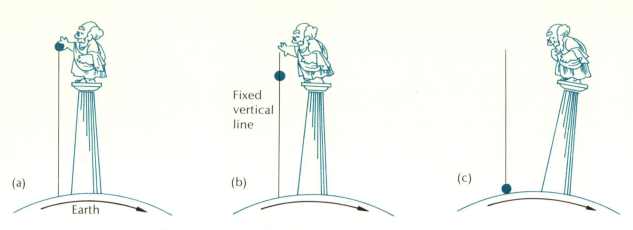

Fixed
vertical
line

(a)

Earth

(b)

(c)

2-1 Aristotle argued that if the earth rotated, a stone dropped from a tower (a) would fall vertically straight down while the tower rotated away from the vertical line of fall (b). When the stone hit the ground (c), it would land considerably west of the base of the tower.

each science must have its own method. In physics, for example, the study of motion observed on earth (local motion) was not related in any way to the study of the motions of stars and planets; further, the mathematical methods in astronomy were not to be used to describe earthly motion. This artificial compartmentalization continued for 2000 years and actually hindered man's search for the unity in nature that Aristotle himself sought.

Aristotle's cosmological view began with the earth located at the center of the universe. He maintained that the earth was at rest and, even today, his arguments for this show "common sense" and are "logical" in view of our everyday sense experiences. The earth was at rest, Aristotle argued, because a stone dropped from a high tower landed directly at its base. If the earth rotated on its axis, it would have moved eastward while the stone was falling (see Figure 2-1); the stone would then have landed a good deal west of the base of the tower, contrary to observation. Aristotle's second argument for the earth being at rest was somewhat weaker, since it was based on the prior assumption that the only natural motion of the element earth was down—in a straight line toward the center of both the universe and the earth. He argued that any circular motion of earth around its center must be forced (not natural) motion and that unnatural, forced motion could not be eternal. Since the order of the world was eternal, it was impossible for the earth to be in motion.

Aristotle correctly reasoned that the earth was spherical in shape. He gave several reasons for this:

> *Its shape must be spherical. For every one of its parts has weight until it reaches the center, and thus when a smaller part is pressed*

on by a larger, it cannot surge round it, but each is packed close to, and combines with, the other until they reach the center. . . . If then the earth has come into being, this must have been the manner of its generation, and it must have grown in the form of a sphere; if on the other hand it is ungenerated and everlasting, it must be the same as it would have been had it developed as the result of a process. . . . Further proof is obtained from the evidence of the senses. (i) If the earth were not spherical, eclipses of the moon would not exhibit segments of the shape which they do. As it is, in its monthly phases the moon takes on all varieties of shape—straight-edged, gibbous and concave—but in eclipses the boundary is always convex. Thus if the eclipses are due to the interposition of the earth, the shape must be caused by its circumference, and the earth must be spherical. (ii) Observation of the stars also shows not only that the earth is spherical but that it is of no great size, since a small change of position on our part southward or northward visibly alters the circle of the horizon, so that the stars above our heads change their position considerably, and we do not see the same stars as we move to the North or South.[1]

Notice that Aristotle was led to the correct idea here by using the evidence of his senses, whereas earlier his sense data had caused him to argue for a motionless earth. Our "common-sense" ideas about nature come partly from our own observations and partly from the culture, language, and teaching with which we grow up and which we use to organize our sense observations. Some of these simple, common-sense ideas are right; Aristotle's concept of a spherical earth is an example. But other ideas are too complex to fit one simple, common-sense picture, as Aristotle's explanation for a motionless earth was. We will see later, with Galileo, how the complexity of a situation can be broken down by designing and performing experiments to test each part of a complex theory by itself.

With a motionless spherical earth at the center of the universe, we proceed to Aristotle's ideas on the composition of everything on earth and up to the moon. In his sublunar region, there were four elements, fire, air, earth, and water, of which everything was composed. These elements were not truly elementary in the sense of Democritus' atoms, but were themselves organized from "prime matter" by the four qualities hot, dry, cold, and moist. Thus fire was hot and dry; water was cold and moist; and so on, and the qualities could rearrange the elements to make things the way they were. Such a distinction between qualities and elements is interesting, but not as important to us as Aristotle's ideas on motion.

[1] Milton K. Munitz, *Theories of the Universe; From Babylonian Myth to Modern Science* (New York: Macmillan, 1965). Used by permission.

ARISTOTLE

Probably no man in the history of thought has had as much influence as Aristotle—the Greek philosopher, political thinker, moralist, educator, biologist, and physicist. Although best known for his work in ethics and philosophy, Aristotle also made contributions to the "hard" sciences, particularly biology, physics, and astronomy. ☐ He was incredibly hard working and developed an interesting method of staying awake to gain study time. Diogenes Laertius, one of his biographers, explains that Aristotle put a big brass basin beside him while he worked and held a weight in his hand over the basin. When he dozed, his hand would relax and drop the weight, producing a noise that was loud enough to wake him. This apparently kept him up (and may have stirred his thoughts on falling objects besides). ☐ In biology, Aristotle made very careful observations from which he developed animal classifications that are still in use today. In physics, however, his methods turned out to be incorrect; among other things, he suggested that bodies fell at a rate determined by their weight, an opinion which Galileo and others corrected 1800 years later. It is astonishing that Aristotle arrived at his conclusion when the fact that bodies of unequal weights fall at nearly the same speed is so easily observable (even without a leaning tower). ☐ Convinced that the universe was basically an organized system, Aristotle was also very interested in organizing systems of thought. He began the process of dividing knowledge into departments—a process still popular today, even when irrelevant—and established a school structured to place major emphasis on informal discussions. ☐ Aristotle also tutored privately, and one of his jobs got him into trouble in a roundabout way. He was hired by King Philip of Macedonia to tutor Crown Prince Alexander, who was later to become Alexander the Great. Long after Aristotle had left the palace, the Athenians revolted against the Macedonians and, recalling his connections with them, accused Aristotle of impiety towards the gods (a popular charge throughout the history of science). Threatened with death, he left Athens, only to die of natural causes a year later at 62.

"Aristotle Contemplating the Bust of Homer" by Rembrandt. Detail.

Metropolitan Museum of Art

Each of the four elements had a "natural motion." Fire and air had "levity" in that they tended to move upward, away from the center of the universe (fire more so than air); water and earth had "gravity" in that they tended to move downward toward the center of the universe (earth more so than water). Aristotle's fundamental "law" of motion was merely the qualitative statement that, except for these natural motions, all other motions of the four elements were forced, unnatural, and *required a mover*. If an object was moved, somebody or something had to move it (forced motion). When something or somebody ceased to move the object, it would return by natural motion to its proper place in the universe. We will pursue this in greater detail when we deal with Aristotle's ideas on projectile motion in the next section.

From Aristotle's fundamental "law," we see several things. First, he emphasized that in the description of motion it was the *end point* of motion that was relevant. He explained the increase in speed (which we now call **acceleration**) of a falling stone by noting that it was approaching its proper place at the center of the universe like a traveler who hurried a bit as he approached his home. We may be amused by this anthropomorphic view now, but the idea that the description of motion depended on the distance of the object from its particular proper place in the universe was so well instilled in the minds of post-Aristotelians that, even 20 centuries later, Galileo and others had severe difficulty overthrowing this concept.

A second point implied by Aristotle's fundamental "law" was that the universe was not a void filled with atoms (Democritus' theory). Rather it was a plenum; it was filled everywhere. Below the moon there were the four elements with their natural motions and with forced motions if a mover intervened. There was no void. From the moon up to the sphere of the stars, there was no void either. These celestial spheres were composed of a fifth element, Plato's ether or **quintessence**, not present on earth. Unlike the motion of the other four elements, the natural motion of quintessence was circular. This circular motion carried the moon and other celestial bodies in their daily travels around the earth. If the moon, for example, were made of one or more of the four terrestrial elements, it would either have to fall in toward the earth (like water or earth) or move away from the earth (like fire and air). Since it was not observed falling toward or moving away from the earth, a further explanation of its circular motion was required—one that was not inconsistent with the physics of fire, earth, air, and water.

Plato and the Pythagoreans taught that a circle was the perfect plane figure; hence circular motion was perfect. There was no forced motion in the celestial region, so it was correct to use mathematics to describe these perfect motions in Aristotle's astronomy. Beyond the spheres of the stars, beyond the confines of the closed finite universe, there was nothing else—neither places, nor void, nor time—for these would presuppose the existence of a natural body different from the universe and lead to a contradiction. Since nature had a purpose, it made no sense to conceive a void beyond the universe because such a concept led to no purpose.

Many people ascribe Aristotle's errors in physics and astronomy to the fact that he used "armchair" arguments and not direct observations. It is true that he did not use observation for the sake of what we now call experiment, but Aristotle was, in fact, a careful observer. Indeed he may even have been *so close* an observer that he was not sufficiently abstract. Consequently, his "laws" or generalizations were too simple to explain some of the more complex phenomena. Yet the mere simplicity of his explanations of natural phenomena was probably the factor most responsible for the immediate acceptance of his theories as well as for the use of Aristotelian physics for almost 2000 years.

2-4
Projectile Motion

Aristotle's ideas on projectile motion are particularly interesting, since they began a long, tortuous chain of events which finally ended with Galileo and Newton. The question is simple: If a projectile (stone) is thrown upward, why does it move *after* it leaves the hand of the thrower? What moves it when it is going up? Clearly, as we have already seen, the projectile's *natural* motion should be down, so that the upward motion is forced or violent motion. Forced motion must be caused by some agent, but, since the projectile is no longer in contact with the thrower's hand, there must be some other agent or element that causes the upward motion. Common sense suggests that the surrounding air may be the agent.

Aristotle considered two theories. The first (see Figure 2-2) contended that the air in front of the projectile moved around in

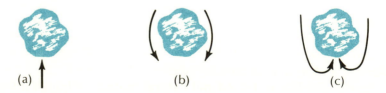

(a) (b) (c)

2-2 According to Aristotle, a stone continues to rise after being thrown up (a) because air rushes in behind it (b) and pushes the stone up (c).

back of it and pushed the projectile upward. Even Aristotle was not satisfied with this theory and claimed that only some people *said* this was an explanation.

The second theory was a slight improvement. The hand moved some air when the object was thrown up. This air then moved another portion of air above it, and the power to move further higher portions of air was conveyed by contact in this way. The stone was then drawn along by the motion of the air surrounding it. However, most of Aristotle's followers were not happy with this explanation either, as we will soon see.

If Aristotle's ideas on projectile motion sound a bit far-fetched, his theory on the speed of natural motion (freely-falling objects) is even worse because it predicts results which are contrary to actual observations. Aristotle contended that the velocity of the object should depend on the object's power to move divided by the resistance afforded by the medium through which the object moved. Thus a heavy stone, which has more "power to move" than a light stone, should fall faster:

> A given weight moves a given distance in a given time; a weight which is as great and more moves the same distance in a less time, the times being in inverse proportion to the weights. For instance, if one weight is twice another, it will take half as long over a given movement.

Another passage from Aristotle's *Physics* that argues for the faster motion of heavy objects relative to light objects is:

> We see that bodies which have a greater impulse either of weight or of lightness, if they are alike in all other respects [shape, size], move faster over an equal space, and in the ratio which their magnitudes [of weight or lightness] bear to each other. Therefore they will also move through the void with this ratio of speed. But that is impossible; for why should one move faster? . . . Therefore all will possess equal velocity. But this is impossible.

We see here again that heavy objects should move faster than light ones. Further, we see another argument against Democritus' void: Where there is nothing to "cleave" (i.e., a void), then there is no resistance to motion and all bodies move with the same velocity which, according to Aristotle's sense experiences, is clearly impossible. Therefore a void cannot exist. Two thousand years later Galileo used a similar argument, but he turned it around to predict correctly that all bodies would fall with the same velocities in a vacuum or void.

Aristotle's ideas of forced and natural motion, although influential, were discussed critically—and, at times, severely—by many post-Aristotelians, but his compartmentalization of each science with its own methods and his concept of a purposefulness in which objects had "natural" places and "natural" motions were less widely contested and much more difficult to change. Aristotle's irresistible, common-sense generalizations based on everyday observations formed a way of thinking that was not to be overcome for many centuries.

2-5

The Late Greek Period

A third period in Greek physics followed immediately after Aristotle, but lasted only a very short time. Archimedes (287–212 B.C.), typical of this period, designed and carried out many experiments in mechanics and fluids and invented a variety of devices of a practical or applied nature, often for use in warfare. He also devised experiments with which he could test the theories he proposed. Unfortunately Archimedes' influence, as well as the experimental method he invented, was all but lost and forgotten until the seventeenth century.

Both Heraclides of Pontus, a student of Plato, and Aristarchus of Samos proposed a simple solution for the observed daily progression of the stars, sun, and planets across the heavens. The phenomena could be accounted for if, instead of the heavenly orbs, the earth rotated once each day about its axis and also revolved about the sun once a year. This was a revolutionary idea, certainly contrary to Aristotelian cosmology, and was even debunked by Aristarchus' contemporary, Archimedes. The principal objection was that if the earth moved in its orbit around the sun, the observed positions of stars seen from the earth should change from one side of the orbit to the other, but they did not. Of course they did, but this effect, called **stellar parallax** (shown in Figure 2-3), could not be observed at the time both due to the enormously large distances of the stars from the earth and the lack of precision equipment. Stellar parallax was not finally measured until the

2-3 Stellar parallax is the apparent change in position of two stars A and B as viewed from the earth at two different positions along the earth's orbit. In January, an observer on the earth sees that A is located to the right of B, while on the other side of the orbit, in June, A appears to the left of B.

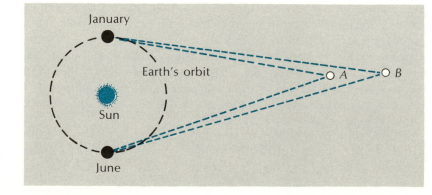

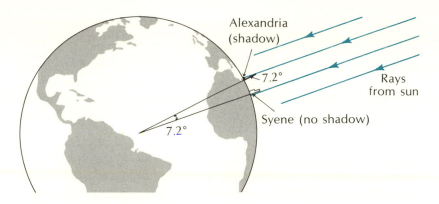

2-4 Eratosthenes' experiment to measure the diameter of the earth.

nineteenth century. Such tremendous distances were evidently incomprehensible to the Greeks, and consequently the sun-centered (**heliocentric**) model of Heraclides and Aristarchus did not gain acceptance at the time.

Another contemporary of Archimedes was the physicist (and geographer) Eratosthenes. Aristotle had made several cogent arguments for a spherical earth, and Eratosthenes set out to measure its circumference. It was known that Syene, a town in Egypt, lay due south of Alexandria; therefore the same great circle or longitude on the earth would pass through both Alexandria and Syene. It was also known that Syene lay on the Tropic of Cancer so that, at the summer solstice (June 21), the sun in Syene would appear *directly overhead* at noon and throw no shadow, while about 500 miles north of Syene, at Alexandria, the sun would cast a shadow at noon. By measuring the angle the sun made on a sundial at Alexandria (7.2°), Eratosthenes used simple plane geometry to show that the same angle was subtended from the center of the earth to the arc from Syene to Alexandria (see Figure 2-4). Since 7.2° is $\frac{1}{50}$ of a complete circle of 360°, Eratosthenes concluded that the circumference of the earth must be 50 times the distance from Syene to Alexandria, about 250,000 stades or 23,300 miles. His result, remarkably close to the modern value of about 25,000 miles, is astonishingly good in view of the errors and inaccuracies in early measurement. The Greeks not only knew that the world was spherical; they also knew its correct size.

This last period of Greek physics was especially important in that experiments began to be used to test theories. However, for reasons that are not particularly clear, the work of Archimedes, Aristarchus, and Eratosthenes had little influence on the future; their ideas and methods were completely overshadowed by Aristotle's influence during the Dark and Middle Ages.

2-6

From Aristotle to Copernicus: Ptolemy and Astronomy

The sciences of physics and astronomy were not completely dead in the centuries after Aristotle. Due to the need for accurate star positions in navigation, the astronomy begun by Eudoxus was even more accurately developed by Hipparchus (140 B.C.) and Ptolemy (120 A.D.). The model described by Ptolemy in his *Almagest* is based on an earth-centered (geocentric) universe with the moon, sun, planets, and stars rotating around the earth in *circular* orbits and with *constant* **angular velocity**. By constant angular velocity, we mean that a celestial body (as observed from the earth) moves through equal angles or arcs in equal times. Thus if a planet traverses one degree of arc along its orbit in ten seconds, it must move through one degree in *any* ten-second period around its orbit (see Figure 2-5).

But it was known from observation that actual planetary motion was not this simple and regular. Sometimes the planets moved faster and sometimes slower, depending on where the planet was on its orbit. Furthermore, the planets were not observed to travel around the earth in perfect circles, but traversed oval-shaped paths. Another striking observation that required explanation was the **retrograde motion** of planets. The planets normally revolve around the earth, traveling across the sky from east to west each day like the moon, sun, and stars. However, they do not revolve at the same rate as the stellar sphere, which contains the distant stars and revolves around the stationary earth once each day. The planetary spheres revolve slightly more slowly than the stars and hence the position of each planet, relative to the background of stellar constellations, changes from day to day, in agreement with our observation.

If we ignore for the moment the daily rotation of all the celes-

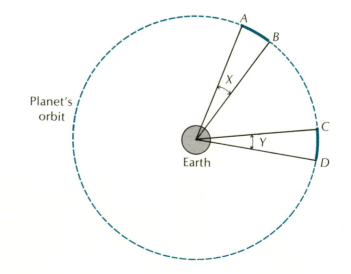

2-5 Constant angular velocity for planetary motion. If it takes a certain amount of time for a planet to go from A to B and the angle it sweeps out is x, then, in the same amount of time, it will sweep out the same angle anywhere along the orbit. If the time it takes to go from C to D is equal to the time it takes to go from A to B, then the angle y equals the angle x.

tial spheres for the moon, planets, sun, and stars, then we observe that the planets slowly move from west to east across the various constellations of stars (which are now assumed motionless). However, at certain regular periods, this motion relative to the stars reverses and a planet moves backward from east to west for several months. Subsequently, the planet again reverses direction and resumes the west to east motion relative to the stellar constellations.

The planet's rotation at nonconstant angular velocity, the noncircular orbits, and the strange phenomenon of retrograde motion all could be explained in the Ptolemaic model (using constant angular velocity and perfect circles) by the use of an additional construct called the **epicycle**. (Ptolemy also used other mechanistic constructs such as the equant and the eccentric to refine his model, but they will not be pursued here.)

The general epicycle arrangement is shown in Figure 2-6. The planet P rotates at constant angular velocity around the small circle (epicycle) centered at C. At the same time, the epicycle's center C rotates around the earth in a circle at a different constant angular velocity. Depending on the velocity of P around C relative to that of C around the earth, a large number of different planetary orbits can be constructed.

In Figure 2-7, for example, an oval-shaped orbit is constructed. The planet rotates around C at exactly the same angular velocity (but in the opposite direction) as C rotates around the earth. Thus when C is $\frac{1}{4}$ of the way around at C', the planet P is $\frac{1}{4}$ of the way around its epicycle at P'; when C is $\frac{1}{2}$ of the way around at C'', the planet is at P''. It is easy to see that the oval-shaped orbit shown in color results from the two combined motions.

Figure 2-8 shows how the epicycle can be used to construct retrograde motion. The angular motion of P around C is a good deal faster than that of C around the earth, and both circles rotate in the same direction (clockwise, as shown in the figure). The planet P can be observed moving clockwise most of the time across the stationary background of stellar constellations. However for a short time near P', the planet stops and moves in the opposite direction as long as the velocity of P', as seen from the earth, is larger than that of C'. Shortly thereafter the planet resumes its normal clockwise motion across the star field. The colored line shows this combined motion as seen from the earth. The arrows indicate the direction the planet is moving in relation to the stars. The complete daily circular motion observed for the stars and planets can finally be obtained from this picture by allowing *both* the circles for the stars and the planetary circle to rotate about the earth together in such a way that they complete their orbits in one day.

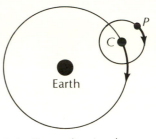

2-6 General epicycle construction.

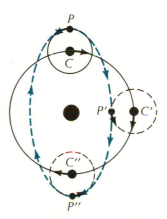

2-7 Use of one epicycle to construct an oval orbit.

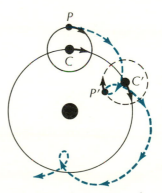

2-8 Use of one epicycle to construct retrograde motion.

PTOLEMY

Claudius Ptolemy was the Greek astronomer and geographer who perfected an astronomical system with a motionless earth at its center. He lived in Alexandria most of his life (about 100 to 170 A.D.); there he compiled a thirteen-volume book, *The Almagest*, in which he expounded his theories. □ Ptolemy was also a cartographer, although his work was adversely affected, as were all early maps, by the lack of accurate instruments. Much of his work was based on a very small diameter for the earth as well, so his calculations were far afield. □ Although his system for classfying stars (six orders of magnitude based on visibility from earth) is still used today, the rest of Ptolemy's theories have been discarded. Many of his calculations were based on stellar and planetary observations made by Hipparchus, a predecessor of Ptolemy's by some 250 years. □ Ptolemy's work greatly influenced Arabic astronomers, who translated the entire *Almagest*. His theory on planetary motion, which employed a complex system of epicycles to explain observable phenomena, was the accepted explanation of the motion of the planets until the seventeenth century.

Alinari-Giraudon

Claudius Ptolemy; relief by Giotto and Andrea Pisano from the bell tower of the Cathedral in Florence.

Thus we see Ptolemy's system as a kind of complicated celestial clockwork which could predict the intricate motions of stars and planets. It was completely founded on the basic premises laid down by Aristotle: a geocentric universe in which all motions of the heavenly bodies were circular (i.e., perfect) and had constant angular velocities (the natural motion of the quintessence). The Ptolemaic system was the first detailed quantitative model of the universe and gained wide acceptance due to its success in predicting planet and star positions.

2-7

Terrestrial Motion

We have seen that Aristotle's model of a geocentric universe (as refined by Ptolemy) was enormously successful in predicting planetary motion. But his terrestrial "laws" of motion governing freely-falling bodies and projectile motion did not achieve the same success. In fact, the continual questioning of these "laws" by many scientists over the centuries eventually led to the discovery of the new physics which was, in time, to change man's outlook on both terrestrial and celestial phenomena completely.

In the sixth century A.D., John Philoponus, a Greek commentator on Aristotle's physics, stated that the "laws" on falling bodies were false. In projectile motion, Philoponus argued that some incorporeal motive force was imparted by the projector to the projectile and that the air had little or no effect on the motion. "Impressed virtue," he believed, remained in the projectile for a short time and was self-expending. It faded away as the projectile became more and more separated from the projector, and then natural motion (gravity) took over. This theory of **impressed virtue** was to be revived at the University of Paris in the thirteenth and fourteenth centuries.

Yet another theory of projectile motion was advanced by John Buridan in the fourteenth century. His "impetus" theory comes closest to a qualitative description of our modern views on motion. **Impetus**, he stated, was what made a projectile go upward, and the amount of impetus depended upon the swiftness of throwing, as well as upon the quantity of matter in the projectile.

By the end of medieval times and the beginning of the Renaissance, many people, including Leonardo da Vinci, were questioning Aristotelian physics and trying to discover more reasonable and more correct "laws" of motion governing bodies near the earth. The stage was being set for the beginning of the first revolution—the Mechanical Revolution. It started in the 1500's and, when it ended with Isaac Newton in 1687, it had overthrown the Aristotelian picture of the universe and established a new science and a new mechanistic philosophy. Although the rumblings of this revolution were heard in the Dark and Middle Ages, the first shot was fired in 1543 when Nicolas Copernicus published *De Orbium Coelestium Revolutionibus.*

Questions

1 According to Aristotle's theory, what role did the five elements play in the motion of objects and what was the important concept in explaining motion? What were the two kinds of motion?

2 Did the Aristotelian theory of motion impose any limitations on the description of motion? Explain.

3 Why was the field of astronomy thought to be of a different nature than that of terrestrial physics? Can you think of any connections between these two fields that Aristotle might have made?

4 Why did the ancients think that mathematics was applicable to astronomy, but not to terrestrial motion?

5 What evidence do we have today for a spherical earth?

6 Can you think of other ways to measure the earth's circumference besides Eratosthenes' method?

7 Aristotle believed in an organismic universe with purposeful ends. In what ways do you find this view satisfying? Dissatisfying?

8 What "common-sense" ideas of natural phenomena that you had as a child changed as you grew older?

9 Do you think that most people prefer simple, "common-sense" explanations for the causes of natural processes rather than more complex ones? Why?

10 Explain Aristotle's arguments against the existence of a void.

Questions Requiring Outside Reading

1 Posidonius (*c.* 100 B.C.) used a method different from Eratosthenes' to measure the circumference of the earth. Describe it.

2 How did Aristotle's compartmentalization of disciplines help or hinder other fields of thought?

3 Describe some of Archimedes' specific contributions to physics.

4 Explain the eccentric and the equant used by Ptolemy.

5 Can retrograde motion be observed in Mercury and Venus, both closer to the sun than the earth? Explain.

Copernicus: The Beginning of the Mechanical Revolution

The first major break with Aristotelian concepts in physics and astronomy was made by Nicolas Copernicus, whose work on a new model of the planetary system was published in 1543, the year of his death at 70. Most of the ideas in *De Revolutionibus* had appeared earlier in unpublished manuscripts which were circulated to various friends who eventually urged Copernicus to publish the entire book. It is understandable why he waited as long as he did to do so; there were great risks involved in challenging the Catholic Church's authority on cosmological matters.

Copernicus challenged the geocentric universe of Aristotle and Ptolemy. According to his new theory, the earth was no longer at the center of the universe, nor did the planets, sun, and stars rotate around the earth. Rather, *the sun was at the center* and the earth, planets, and stars revolved about it. Furthermore, the earth was not stationary, as Aristotle would have had us believe, but rotated on its axis once every day. Thus the stars, which could be thought of as stationary or fixed, only *appeared* to rotate about the earth. Think of the impact this theory must have had on the scientists and laymen of that time. It shook to the very foundation the ingrained traditional thought of all philosophers, theologians, and scientists and, as a result, Copernicus' ideas could not be accepted. But the theory had a definite impact on the young imaginative minds of a few people like Bruno, Galileo, and Kepler, who saw clearly the enormous advantages of the Copernican system over the established Ptolemaic-Aristotelian model of the universe.

3-1

Introduction

COPERNICUS

When Nicolas Copernicus was born in Poland in 1473, the world had not moved since 300 B.C. when Aristotle had explained that a stationary earth existed as the center of the universe. But Copernicus described a new universe which excited a few men who had vision, imagination, and courage. These Copernicans were to lead the way to a new physics, and with this new physics was to come a revolution in western thought, culture, and politics. ☐ His father died when Copernicus was ten, and he and the other children in the family were cared for by their uncle. Fortunately for Nicolas, his uncle was on his way to becoming an important Catholic bishop—a position rich with power, influence, and favors. After attending the University of Cracow and traveling and studying in Italy, Copernicus returned home to find the position of canon at Varma open to him. A canon was a church functionary with very little function. Although he was supposed to remain at the church to which he was attached, even that was not a strict requirement. Canons were supported by taxes and had a great deal of freedom to do as they wished, so the job did have its advantages. ☐ Copernicus spent some of his time as an economist (his theory on bad money driving out good is still held) and he was a student of the classics and an author of Latin prose. But he took his canonical position seriously enough to serve as a doctor for the poor at no charge as well as a general when church property was attacked by the Knights of the Teutonic Order. ☐ He began his work in astronomy in about 1506 and continued it for the rest of his life. His theory—based on the idea of a moving earth—was first published as a monograph and circulated in 1514, but it was too general and philosophical to cause any great upheaval. It was a theory similar to others that had been advanced earlier by various Greek philosophers including the Pythagoreans. Copernicus subsequently began the important task of compiling data to support his theories, since he realized that theory could only be justified by good observations. ☐ He worked for 20 years on his second book, *De Revolutionibus Orbium Caelestium,* avoiding publication partly because the Inquisition had begun executing people with unorthodox ideas and partly because he was dissatisfied with parts of the book. However, when a young Lutheran professor named Rheticus heard about Copernicus and went to visit him, he was struck so by the simplicity and beauty of the Copernican system that, after ten weeks of working with Copernicus, Rheticus wrote *Narratio Prima,* an analysis of Copernicus' work. The book aroused enough public interest to induce Copernicus to publish his own work. ☐ The astronomer was never completely satisfied with this book—he never had enough time to finish it the way he wanted to—but Rheticus took it to Nuremburg and had it printed. The first copy was given to Copernicus on May 24, 1543, just a few hours before he died. Much later it was placed on the index of forbidden books, where it remained until 1835 when the Catholic Church finally decided that the earth was, perhaps, in motion.

3-2
Retrograde Motion

One major advantage of the Copernican model was its explanation of planetary retrograde motion. This peculiar phenomenon, previously explained satisfactorily by the epicycle construct, arose *without the use of epicycles* as a *natural* consequence of the Copernican theory.

Because of the earth's rotation on its axis (**diurnal motion**), the sun, stars, and planets all rise and set in the sky during a 24-hour day. However the earth as well as the planets also revolves around the sun. Thus we observe the position of a planet relative to the background of "fixed" stars in constant change from day to day.

In Figure 3-1, a planet moves across the various constellations in the same direction (clockwise) until the distance between the planet and the earth is at a minimum. At this point the angular motion of the earth relative to the planet is largest, and the planet, when superimposed on the fixed stars, *appears* to be moving backward (counterclockwise) from May to July. As time progresses and the earth-planet separation increases again, the planetary motion relative to the stars resumes its normal direction.

Copernicus' explanation of retrograde motion was so appealing in its simplicity that it was adopted by quite a few mathematicians and physicist-philosophers. However his heliocentric theory was not generally accepted and actually was bitterly fought for several reasons. One of these, based on theological grounds, was expressed by Luther and Calvin who both believed in the literal interpretations of those passages in the Bible which clearly disagreed with the Copernican theory. Yet the Catholic Church did not officially disagree with the Copernican view at the time. In fact, it gave the "Imprimatur" to Copernicus' book which was, incidentally, dedicated to Pope Paul III. It was not until much later that the church, perhaps forced by Galileo, took an uncompromising position against the Copernican theory.

Nicolas Copernicus

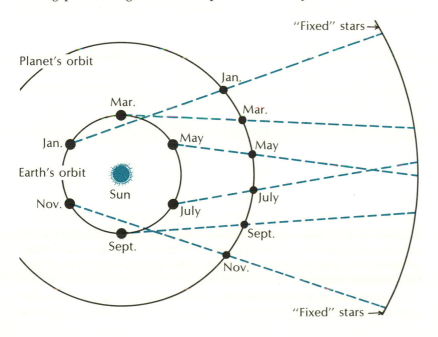

3-1 Copernicus explained retrograde motion without the use of epicycles.

3-3

Copernicus versus Ptolemy

Regardless of the theological issues, the Copernican theory had one major drawback that, even in the time of the early Greek scientists, eliminated any serious consideration of the earth's rotation about the sun—the lack of any observable stellar parallax. If the stars were not all the same distance from the earth then the yearly rotation of the earth around the sun should produce a noticeable shift in the position of a close star relative to the background of the more distant stars (see Figure 2-3, page 16). Copernicus explained that the lack of any observable parallax was due to the immense distance of all the stars from the solar system. Thus the shift would be too small to detect. But the required distances were much too great for his critics to comprehend, so they would not accept his theory scientifically. It was not until several hundred years later in 1838 that refined astronomical observations were finally able to detect stellar parallax and this objection to the theory could finally be removed.

Copernicus still preserved intact the basic Aristotelian premises of circular motion and constant angular velocity. Thus he, like Ptolemy, had to resort to epicycles to explain the variations in solar, lunar, and planetary distances from the earth (or, in his case, from the sun), and to explain the corresponding variations in angular velocities. This was due in part to his use of Aristotle's "perfect" circles and in part to some earlier inaccurate astronomical data. In actuality, Copernicus' model was not that much simpler in the calculation of stellar and planetary positions than Ptolemy's. The enormous simplification in the former model was that epicycles were no longer required to explain retrograde motion, which reduced the number of arbitrary constructs necessary to explain the data.

But how can two different theories such as those of Ptolemy and Copernicus both satisfactorily explain the same astronomical observations? Surely one must be "right" and the other "wrong," but is it possible to decide in any ultimate way which is which? Or is it just a matter of taste and personal preference, as it may be in deciding between two economic theories?

If both theories explain the existing observations with equal accuracy, then it would seem sensible to try and obtain new data or devise a new experiment that hopefully would differentiate between the two. If such a definitive experiment is not possible, we must then rely on some other criterion to resolve the dilemma. Rival theories have existed in physics from the time of Copernicus down to the present day, and not all of these can be judged right or wrong by a definitive experiment. We learn from the eventual acceptance of the Copernican theory that a new way to judge a theory is on the basis of the simplicity and economy of the constructs on which it is based.

However, the situation was not quite this simple. There were other subtle criteria for the acceptance or rejection of the Copernican picture. As we have already learned, Aristotle separated the disciplines of physics (local motion on the earth) from astronomy and cosmology. This may seem at first to be reasonable. But in spite of this separation, there were connections between the two fields that complicated the question of whether or not Copernicus' theory was as "good" or "better" than the Aristotelian-Ptolemaic theory.

The complications arose when the motion of the earth was considered. In the old theory, the earth was motionless, while in the Copernican theory it rotated on its axis once a day and revolved about the sun once each year. If the new sun-centered theory of the solar system was to be believed, an explanation for the motion of objects on the earth must be provided. And here is the connection between the two fields. For *if* the earth were at rest, then the Aristotelian "laws" of physics would explain the upward and downward motion of objects on the earth. But if the earth rotated once each day (a peripheral speed at the equator of about 1000 mi/hr), clearly Aristotle's "laws" of physics could not explain the local motion of objects on the earth. A different set of "laws" was needed.

Thus if the Copernican theory was to be accepted and the Aristotelian-Ptolemaic astronomical system abandoned, Aristotelian physics must also be forsaken and a new physics formulated which would explain the upward and downward (local) motion of objects on the surface of the earth which is itself moving with such great speed. The dilemma was not easy to overcome; the reasons for a stationary earth given by Ptolemy were very difficult to answer:

> In the same way as before it can be proved that the earth cannot make any movement whatever in the aforesaid oblique direction, or ever change its position at all from its place at the center.[1]

Ptolemy argued that the earth and the objects (or weights) on it could not have a common motion (move at the same speed). Thus the earth could only be at rest:

> But, of course, if as a whole it had had a common motion, one and the same with that of the weights, it would, as it was carried down, have got ahead of every other falling body, in virtue of its enormous excess of size, and the animals and all separate weights would have been left behind floating on the air, while the earth, for its part, at

3-4

The Need for a New Physics

[1] This and the following Ptolemy excerpts on page 28 are from T. L. Heath (trans.), *Greek Astronomy* (London: J. M. Dent & Sons, Ltd., 1932). Used by permission.

its great speed, would have fallen completely out of the universe itself. But indeed this sort of suggestion has only to be thought of in order to be seen to be utterly ridiculous.

Next, Ptolemy attempted to refute Aristarchus' scheme of an earth rotating daily on its axis:

Certain thinkers have concocted a scheme which they consider more acceptable, and they think that no evidence can be brought against them if they suggest for the sake of argument that the heaven is motionless, but that the earth rotates about one and the same axis from west to east, completing one revolution approximately every day. These persons forget however that, while, so far as appearances in the stellar world are concerned, there might, perhaps, be no objection to this theory in the simpler form, yet, to judge by the conditions affecting ourselves and those in the air about us, such a hypothesis must be seen to be quite ridiculous. . . . yet they must admit that the rotation of the earth would be more violent than any whatever of the movements which take place about it, if it made in such a short time such a colossal turn back to the same position again, that everything not actually standing on the earth must have seemed to make one and the same movement always in the contrary sense to the earth.

For, even if they should maintain that the air is carried round with the earth in the same way and at the same speed, nevertheless the solid bodies in it would always have appeared to be left behind in the motion of the earth and air together, or, even if the solid bodies themselves were, so to speak, attached to the air and carried round with it, they could no longer have appeared either to move forwards or to be left behind, but would always have seemed to stand still, and never, even when flying or being thrown, to make any excursion or change their position, although we so clearly see all these things happening.

It was only with great difficulty and much labor that the seventeenth century physicists who believed in the Copernican theory were finally able to refute Ptolemy's objections to a moving earth. Copernicus himself attempted to disprove these age-old arguments against the earth's motion and, in so doing, he laid the foundations on which Galileo, Descartes, and Newton were later to build the new physics. We see how this was accomplished in the following quote from Copernicus' *De Revolutionibus*:

It is claimed that the earth is at rest in the center of the universe and that this is undoubtedly true. But one who believes that the earth rotates will also certainly be of the opinion that this motion is natural and not violent. Whatever is in accordance with nature produces effects which are the opposite of what happens through violence. Things upon which violence or an external force is exerted must become annihilated and cannot long exist. But whatever happens in the course of nature remains in good condition and in its best arrangement.

This idea that the earth's rotation was not violent, but natural represented the beginning of the concept of **inertia**—that an object, once moving, will continue to move "in the course of nature." Here Copernicus introduced another kind of natural motion. In addition to the up–down natural motions of Aristotle, things on the earth also moved sideways, perpendicular to the gravity or levity direction, as a natural consequence of the rotation of the earth to which they were attached:

> *Thus, as has been claimed, a simple body has a simple motion as long as the simple body remains in its natural position and retains its unity. In this position its motion is merely the circular motion, which being entirely within the body, makes it seem to be at rest.*

Copernicus also alluded to **compound motion**, which was to be developed later in more detail by Galileo, as being comprised of a natural circular motion caused by the earth's rotation and a vertical motion due to gravity or levity:

> *We must admit that the motion of falling and rising objects is, with respect to the universe, a double one, compounded always of rectilinear and circular motions.*

This was an important point in the development of the concept of inertia and in the refutation of Ptolemy. In ascribing rotational motion to the earth rather than the heavens, Copernicus wrote:

> *Without cause, therefore, Ptolemy feared that the earth and all earthly things if set in rotation would be dissolved by the action of nature. But why did he not fear the same, and indeed in much higher degree, for the universe, whose motion would have to be as much more rapid as the heavens are larger than the earth? Or have the heavens become infinite just because they have been removed from the center by the inexpressible force of the motion; while otherwise, if they were at rest, they would collapse?*

With this passage, Copernicus made a definite break with Aristotelian methods. The circular movement of the heavens was no longer an idealized geometrical motion; it was subject to the same "inexpressible force" (**centrifugal force**) that actual earthly objects felt when made to move in a circle. (A stone, whirled around on a string, will pull outward on the string.) Thus Copernicus applied terrestrial sense experiences with circular motion to the rotating celestial spheres and thereby implied that celestial motion obeyed the same physical principles as terrestrial motion—a distinct departure from Aristotle's teachings.

Copernicus went on:

Certainly if this argument were true the extent of the heavens would become infinite. For the more they were driven aloft by the outward inpulse of the motion, the more rapid would the motion become because of the ever-increasing circle which it would have to describe in the space of twenty-four hours; and, conversely, if the motion increased, the immensity of the heavens would also increase. Thus velocity would augment size into infinity, and size, velocity. But according to the physical law that the infinite can neither be traversed, nor can it for any reason have motion, the heavens would, however, of necessity be at rest.

But it is said that outside of the heavens there is no body, nor place, nor empty space, in fact, that nothing at all exists, and that, therefore is no space in which the heavens could expand; then it is really strange that something could be enclosed by nothing. . . . Now, whether the world is finite or infinite, we will leave to the quarrels of the natural philosophers; for us remains the certainty that the earth, contained between poles, is bounded by a spherical surface. Why should we hesitate to grant it a motion, natural and corresponding to its form; rather than assume that the whole universe, whose boundary is not known and cannot be known, moves? And why are we not willing to acknowledge that the appearance of a daily revolution belongs to the heavens, its actuality to the earth?

Copernicus then discussed **relativity** theory, a subject which was to become important in physics in his time and remain so through Einstein's work in the twentieth century. Copernicus argued:

Is it not as correct to assume that the earth rotates on its axis with the sun stationary, as to assume that the sun rotates around the earth?

According to Copernicus, only **relative motion** could be detected:

The relation is similar to that of which Virgil's Aeneas says: 'We sail out of the harbor, and the countries and cities recede.' For when a ship is sailing along quietly, everything which is outside of it will appear to those on board to have a motion corresponding to the movement of the ship, and the voyagers are of the erroneous opinion that they with all that they have with them are at rest. This can without doubt also apply to the motion of the earth and it may appear as if the whole universe were revolving. . . .

Copernicus had several more arguments for a rotating earth. One—based on the idea that the heavens were more noble and more perfect than earth and, therefore, must be at rest—also contained the argument that the rotating earth moved around the sun once each year as well:

Moreover, the condition of rest is considered as nobler and more divine than that of change and inconstancy, so the latter would, therefore, be more suited to the earth than to the universe. . . . It is clear, therefore, from all this, that motion of the earth is more probable than rest, especially in relation to the daily rotation, which is most characteristic of the earth.

Since nothing stands in the way of the movability of the earth, I believe we must now investigate whether it also has several motions, so that it can be considered one of the planets. That it is not the center of all the revolutions is proved by the irregular motions [retrograde motion] of the planets, and their varying distances from the earth, which cannot be explained as concentric circles with the earth at the center. . . . If one admits the motionlessness of the sun, and transfers the annual revolution from the sun to the earth, there would result, in the same manner as actually observed, the rising and setting of the constellations and the fixed stars; and it will thus become apparent that also the haltings and the backward and forward motion of the planets are not motions of these but of the earth, which lends them the appearance of being actual planetary motions. Finally, one will be convinced that the sun itself occupies the center of the universe. And all this is taught us by the law of sequence in which things follow one upon another and the harmony of the universe; that is, if we only (so to speak) look at the matter with both eyes.

It should be pointed out that, at the time, the publication of these ideas was acceptable to the church as long as the Copernican theory was considered a hypothesis: useful for calculation, but not believed in as an absolute truth. It was not until half a century later that the acceptance or teaching of the Copernican doctrine was to become risky. Giordano Bruno, believed born in 1548, was to be one of the first martyrs of physics. He speculated on the existence of worlds other than the earth which, in analogy to the Copernican theory, might form planetary systems around some star:

Sky, universe, all-embracing ether, and immeasurable space alive with movement—all these are of one nature. In space there are countless constellations, suns, and planets; we see only the suns because they give light; the planets remain invisible, for they are small and dark. There are also numberless earths circling around their suns, no worse and no less inhabited than this globe of ours. For no reasonable mind can assume that heavenly bodies which may be far more magnificent than ours would not bear upon them creatures similar or even superior to those upon our human Earth.[2]

[2] American Foundation for Continuing Education, *Exploring the Universe* (New York: McGraw-Hill, 1963), p. 304. Used by permission.

General Research and Humanities Division, New York Public Library: Astor, Lenox, and Tilden Foundations

Frontispiece from
Opere di Giordano Bruno
by Adolfo Wagner.

For this and other heresies against the church involving advocacy of the Copernican system, Bruno was burned at the stake by the Inquisition in 1600—the beginning of the century which was to be called the "Age of Reason."

Even though he died five years before Bruno was born, Copernicus himself was probably fearful of the Inquisition and, for this reason, may have delayed publication of his theory until the last year of his life. In any event, when his painstakingly detailed mathematical calculations of the planetary orbits did appear in *De Revolutionibus*, they bore the following preface (perhaps apology is a better description) by Andreas Osiander, a Lutheran clergyman and friend of Copernicus:

To the Reader Concerning the Hypotheses of This Work
Since the novelty of the hypotheses of this work has already been widely reported, I have no doubt that some learned men have taken serious offence because the book declares that the earth moves, and that the sun is at rest in the center of the universe; these men undoubtedly believe that the liberal arts, established long ago upon a correct basis, should not be thrown into confusion. But if they are willing to examine the matter closely, they will find that the author of this work has done nothing blameworthy. For it is the duty of an astronomer to compose the history of the celestial motions through careful and skillful observation. Then turning to the causes of these motions or hypotheses about them, he must conceive and devise, since he cannot in any way attain to the true causes, such hypotheses as, being assumed, enable the motions to be calculated correctly from the principles of geometry, for the future as well as for the past. The present author has performed both these duties excellently. For these hypotheses need not be true nor even probable; if they provide a calculus consistent with the observations, that alone is sufficient. . . . Now when from time to time there are offered for one and the same motion different hypotheses (as eccentricity and an epicycle for the sun's motion), the astronomer will accept above all others the one which is the easiest to grasp. The philosopher will perhaps rather seek the semblance of the truth. But neither of them will understand or state anything certain, unless it has been divinely revealed to him. Let us therefore permit these new hypotheses to become known together with the ancient hypotheses, which are no more probable; let us do so especially because the new hypotheses are admirable and also simple, and bring with them a huge treasure of very skillful observations. So far as hypotheses are concerned, let no one expect anything certain from astronomy, which cannot furnish it, lest he accept as the truth ideas conceived for another purpose, and depart from this study a greater fool than when he entered it. Farewell.[3]

[3]American Foundation for Continuing Education, *Exploring the Universe* (New York: McGraw-Hill, 1963), p. 111. Used by permission.

BRUNO

Giordano Bruno came from Naples and began his higher education in a Dominican monastery. Monastic life did not agree with him though, especially when he was warned of suspicions that he was a possible heretic. He never knew why he was suspected, but finally left the order and began traveling around the world working in print shops and writing. □ One of his important works was a book on the art of memory from metaphysical and psychological viewpoints. The book was well known and respected, giving him a name at universities and a key into many of the courts in Europe. He also wrote a satiric play, "The Torchbearer," which many critics considered the best comedy of the period. □ Bruno spent a good deal of time in relative peace at the Sorbonne and then went to London where he wrote some of his best philosophical works—one, a book supporting the Copernican theory. He traveled in the best circles and loved the women ("gracious and gentle, soft and tender, young, fair and delicate, blond, white of chin, pink of cheek, of enticing lips, eyes driving, breasts of ivory, and hearts of adamant"). Apparently they liked him, too. □ Becoming homesick, Bruno at last accepted an invitation from a rich Venetian named Mocenigo to teach his art of memory in Venice. The two men did not get along well at all, however. Mocenigo wanted instant power over man and nature, a power Bruno was unable to grant. The church's apparatus, though grinding slowly, had discovered that Bruno, now a "famous heretic," was back in Italy. At the same time, Bruno, realizing that no immediate reconciliation between himself and the church was in sight, decided that he might be safer elsewhere. □ When Bruno said that he was going to Frankfurt to publish a book, Mocenigo had him locked in a cellar, went to the church authorities, and told them that Bruno believed Christ to be a magician—a rather serious heresy. Bruno was turned over to the Inquisition and spent the next nine years in Venice and Rome defending his doubts about such Catholic doctrines as the Immaculate Conception as well as his beliefs that the earth moved and that there might be other inhabited worlds besides our earth in other solar systems. □ He was finally burned at the stake in 1600. When the inquisitors read him the sentence, Bruno's words were, "Greater your fear in pronouncing this sentence than mine in receiving it."

Was Osiander right in saying that the purpose of the Copernican system was merely "to provide a correct basis for calculation?" Or was there something more fundamental involved? Galileo, Descartes, Bruno, and Kepler believed the latter and, at great personal risk, each tried to extend and teach the new system of the universe begun by Copernicus.

The Copernican theory has been called a revolution in itself. Copernicus "invented" one of the two essential ingredients of modern science when he showed that traditional concepts and "laws" do not have to be followed faithfully or merely improved upon to increase our understanding of nature. Instead, tradition must be constantly questioned. If revolutionary new concepts explain the phenomena with *fewer arbitrary assumptions*, they must also be considered. Such simplicity or economy of basic assumptions should not be confused with the simple, common-sense causes for physical phenomena given by Aristotle. Here, the point is that a theory is judged better than a competing one if it requires fewer **initial assumptions** or fewer **basic constructs** to explain the observed phenomena; whether or not the detailed (mathematical) explanation of any specific phenomenon is simple or complicated is not important. For Aristotle, the explanation of each phenomenon had to be simple and understandable via common-sense arguments that followed from a set of several sometimes inconsistent basic constructs. The new view required instead a simple and consistent original construct such that a phenomenon like retrograde motion could be explained *as a natural consequence* of that construct. This point of view will occur again in the works of Galileo, Kepler, Newton, and physicists throughout the years to the present day.

Questions

1 What features of the Ptolemaic cosmology were retained in the Copernican system? What features were changed? In what way was the Copernican model of the universe simpler? Did its simplicity make it better? Why or why not?

2 Explain what Ptolemy meant when he said, "the earth . . . at its great speed, would have fallen completely out of the universe itself." Does this seem reasonable to you?

3 What were some of the revolutionary ideas for which Giordano Bruno was executed by the Inquisition? Are any of his ideas accepted or given credence today?

4 In these days of "publish or perish," scientists cannot afford to wait until their last years as Copernicus did to make their discoveries public. As a result, many incomplete studies are published. Do you think this is a better method? Why or why not?

5 How did Copernicus' ideas on the circular motion of the heavens differ from those of Ptolemy and Aristotle?

6 Is it necessary to assume that there is some order to nature in order to make science meaningful? Would the study of actual chaos be sensible in any way?

7 Do you think that there might be some meaning or intuitive order in the universe which man will never understand? Explain.

8 Are the laws of physics dependent upon our language (i.e., upon the manner in which we describe events)? Do you think they should or should not be?

9 Explain the meaning of Copernicus' statement, "[The body's] motion is merely the circular motion, which being entirely within the body, makes it seem to be at rest."

10 Do you think that science and religion are seeking the same or different truths? Explain.

11 Explain Ptolemy's main objections to a moving earth. How did Copernicus refute them?

12 From your own experience, give some examples of the relativity of motion (for example, how it is sometimes hard to tell whether you are moving past an object or it is moving past you).

13 If the solar system consisted only of the earth, moon, and sun, do you think that the Copernican system would be necessary? Explain.

14 Is there any way we *know* that the stars do *not* rotate once each day around the earth?

Questions Requiring Outside Reading

1 What is Occam's Razor? What was the role of this principle in the development of physics in the Middle Ages and the Renaissance? What was the influence of this principle in the debate concerning the two cosmological systems, Ptolemaic and Copernican?

2 What previously accepted ideas did Copernicus use in his revolutionary model of the universe?

3 What are some of the specific passages from the Bible that dispute the Copernican theory?

4 What were some of the pre-Copernican ideas concerning inertia (i.e., that an object can have a natural motion different from the gravity or levity in the vertical direction)?

5 Did Copernicus believe in an infinite universe? Did he believe in the Aristotelian plenum or in Democritus' void filled with atoms?

6 Compare the Copernican revolution with revolutions that occurred in the sixteenth century in other fields (i.e., in literature, philosophy, art, music, etc.).

4

Galileo and the Problem of Terrestrial Motion

4-1
Introduction

The second major personality in the birth of modern physics was Galileo Galilei (1564–1642). Galileo's contributions to the new physics were enormous due both to the diversity of his interests and to the depth of his insight and creativeness. He was well known as an astronomer: He was the first person to use the telescope for real astronomical observation; as a physicist: He discovered the two fundamental laws of motion for objects on the earth, thereby refuting Aristotle's laws of terrestrial motion; as the "father" of modern scientific methods: He discovered the role of precise, controlled experiments in understanding nature and required that mathematics be used to describe all natural phenomena, not simply celestial motion as Aristotle had taught. Copernicus recognized the beginnings of many of the solutions to the problems of local terrestrial motion on a moving earth. Galileo was able to take a huge step beyond this to find the solutions experimentally and formulate them explicitly in precise mathematical form. In this way, he was able to chop away a large amount of the enormous edifice of Aristotle's scientific methodology, an edifice already doomed by the work of Copernicus.

4-2
Galileo Sees Motion in a New Way

Aristotle's fundamental question regarding motion was, "Why does an object move from its proper place?" This was replaced by Galileo's question, "How does an object *change* its present state of motion (or rest)?" For Aristotle, the distance of an object from the end point—its "proper place" in the universe—was important in describing motion. Galileo, although at first confused by this traditional concept, was able to realize in understanding motion that (1) it was necessary to know the distance from the *starting point* and (2) *time* was the important missing ingredient.

Galileo contributed two further essential foundations to modern physics. First, he insisted that local motion of objects on earth should be described **quantitatively** (with numbers), not qualitatively as before. Questions such as *how much time* and *how far* made mathematics, previously used only in astronomy, an essential tool in physics as well. Galileo stated, "The book of nature is written in mathematical characters." His second contribution was perhaps the most crucial one for all of modern science: teaching the importance of designing and performing a sufficiently accurate experiment to test a proposed theory. In Galileo's time, experimentation was very unorthodox in physics and this major contribution was not easily achieved.

Why was it so difficult? First, we have to remember that much of Aristotle's way of thinking was based on everyday observations and common-sense generalizations. It is hard to convince others that their everyday observations are not simple at all, but are composed of many complicated factors, some not even known or understood. Let us take rolling a ball on a level table as an example. We give the ball a push and it rolls along, slowing down constantly and finally coming to rest. This is our observation and we presume we are in agreement with the other observers who are watching the ball roll.

However all observers do not interpret an observation in the same way. Aristotle felt the fact that the ball slowed down and stopped was only natural: Constrained by the table, it could not exhibit its "natural" vertical motion downward toward the center of the universe. With no natural motion allowed and no mover in contact with it, the ball would seek its "proper place" in the universe—a state of rest.

Galileo, however, arrived at a completely different interpretation of the same situation. The differences in the slowing-down rates of rough balls compared with those of smooth balls suggested to him that the surface of the table offered **resistance** to the motion of the ball; he further perceived that the air also offered resistance to the ball's motion. Although these factors were too complicated for him to analyze in detail, he knew they affected the ball's motion in an important way.

The great imaginative leap here was that Galileo did not analyze the motion *exactly as observed* as Aristotle would have done; instead he **abstracted** only that part of the ball's motion which would be left over if there were no resistance from the air or from the table surface. The imagination and intuition this required are evident in Galileo's description of a third kind of motion which was neither natural nor forced:

> . . . on a smooth horizontal plane any body, meeting with no external resistance, can be moved by as small a force as wished.[1]

We find from Galileo that, in the absence of resistance, a body will move at constant velocity on a horizontal plane.

> . . . along a horizontal plane the motion is uniform since here it experiences neither acceleration nor retardation.[1]

If acting alone, he added, this velocity "would carry the body at a uniform rate to infinity."

So Galileo made statements *not* about what he actually observed, but about what he believed would be observed if air resistance and other factors were neglected. He abstracted. Instead of treating the **decelerated** motion of the ball in its entirety as Aristotle would, Galileo saw that the ball's overall motion was composed of different parts: the original motion (velocity) immediately after the ball was pushed and the subsequent retarding effects of air and table surface resistance. So he dissected these compound effects, threw out the resistance effects on the grounds that they were nonessential to the "true" nature of the motion, and arrived at the abstraction: a theory of motion in the absence of resistance.

But this in itself was not enough. Galileo then went on to support his theory by devising experiments to test specifically the effects of different surfaces and other factors on the ball. This is always the acid test; we will see it used again with Galileo's inclined plane and many other theories throughout the history of physics. If the experimental observations do not conform to or support the theory, then the theory must be discarded or revised. *Even if the experiments agree with the theory, the theory is not "proved;"* however, it can at least be allowed to stand until some contrary evidence appears.

It is interesting to compare Galileo's methods with those of his contemporary René Descartes, who was also working on the problem of terrestrial motion. A first-rate mathematician and philosopher, Descartes employed deductive reasoning. By systematically doubting the real existence of everything by which man is cognitive, he arrived at the *one* thing that could not be doubted—his own consciousness. "*Cogito, ergo sum*"—I think, therefore I am—was the beginning of his philosophy. By pure

The Granger Collection

Galileo Galilei

[1]Laura Fermi and Gilberto Bernadini, *Galileo and the Scientific Revolution.* Copyright © 1961 by Basic Books, Inc., pp. 99–100.

GALILEO

Galileo Galilei's reputation for getting into trouble by challenging orthodox beliefs began early in life. When he was a professor at the University of Pisa, he suggested that two bodies of different weights would fall at the same speed, which directly contradicted Aristotle's teaching. Aristotle's doctrines were supreme in the sixteenth century, so Galileo's theory met with strong opposition. And although there is no evidence that Galileo actually dropped two lead weights off the Tower of Pisa while the university faculty looked on, the blindness of his colleagues was no less grandiose. □ Galileo left Pisa after he was forced to resign from the university there. He accepted a post in Padua in 1592, the same year Bruno was arrested in nearby Venice. By this time, Galileo had become a believer in the Copernican system and began corresponding with Kepler, another young Copernican. Galileo's interest in astronomy grew when he used the telescope, which had been invented somewhat earlier in Holland, for the first time in astronomical observations. By the time he had built his fourth telescope, he had discovered, among other things, that the Milky Way was made up of myriads of individual stars and that there were four moons around Jupiter. □ Galileo suggested that the existence of these four moons meant that the Copernican system was possible, for if Jupiter could revolve around the sun without losing its moons in the process, the earth could certainly do the same. But he was attacked by the Aristotelians on this point. One of them, Sizzi, declared that Galileo had to be wrong because seven was a sacred number (seven metals, seven openings in the head, and so on); therefore there could not be more than seven "celestial orbs." Sizzi and Galileo's other critics refused to look through his telescope. The astronomer characterized these farsighted professors as trying to "argue new planets out of heaven." □ The wrath of the church also began to descend on Galileo's head; he was warned in 1616 by Cardinal Bellarmine, who had been active in Bruno's persecution, to abandon Copernicanism. Galileo, who was sure he had a friend in the Pope, ignored the warning. Unfortunately, Galileo's enemies convinced the Pope that he was one of the characters in a book describing a fictitious dialogue about the universe, and the Pope decided that Galileo must be investigated. He was finally brought before the Inquisition in 1632 where, under threat of torture, he was forced to deny his beliefs. Perhaps feeling that the truth would stand eventually, with or without his death, he denied what he knew to be true—that the earth revolved about the sun. □ He was placed under house arrest for the nine remaining years of his life. During this time, in spite of the fact that he was blind and very sick, he was considered so dangerous that only a few people were allowed to see him. (He was visited by Milton, who wrote a poem about the trip.) Galileo completed one last work, *Two New Sciences,* in which he analyzed his work on motion, acceleration, and gravity. All of his books, like those of Copernicus', remained on the list of unreadable literature for almost 200 years.

reason alone, Descartes argued for existences external to our own consciousness—God, our body, material objects—the sensible world around us.

Descartes proposed a new physics based, like his philosophy, on deductive reasoning: arguing from undeniable (self-evident) truths to arrive at specific predictions for the motion of objects. Although most of his physics turned out to be wrong, in his

Oeuvres (1644), Descartes stated:

> *And the proofs of all this [his incorrect laws of motion] are so certain, that although experience would make us see the contrary, we would nevertheless be obliged to accord more faith to our reason than to our senses.*

How different the approach of this master of "The Age of Reason" is when compared with Galileo! We can begin to see why even Galileo's methods of abstraction, intuition, and experimentation were suspected by his contemporaries. Galileo had to invent and prove the validity of a completely new way of acquiring knowledge—controlled, precise experimentation to enable the observer to isolate the relevant aspects of a problem (like motion) from complicating, irrelevant aspects (like friction).

René Descartes by Gio Batta Weenix, c. 1647.

Collection Central Museum, Utrecht

DESCARTES   René Descartes was born in France in March of 1596. His mother died almost immediately after his birth and he was raised by his grandmother. Like most of the other great minds in history, he was continually asking questions about why things were the way they were.   □   Descartes was sent to a Jesuit school at eight. He always respected the Jesuits, and some of the faculty at the school were to remain his friends throughout his lifetime. Finding the contradictions between the sciences and philosophy frustrating, he turned to mathematics which seemed so clear and conclusive to him.   □   He left school still frustrated and irritated at what he considered to be a good deal of wasted time interspersed with the time well spent. He then spent two years (1612–1613) learning riding and fencing (as every gentleman must) and decided that by using reason to figure out the theory of an art, one could learn it.   □   In 1613, Descartes went to Paris, where he had a wild time for a while until he began studying seriously on his own. He joined a French regiment in Holland to see more of the world and met quite a few mathematicians in the army—one of them, Isaac Beekman.   □   Descartes had stopped on a street in Breda to read a notice of a mathematical problem and asked a stranger (Beekman) to translate it for him. Beekman, a college principal, sarcastically asked Descartes to solve the problem; he did so the next day, much to Beekman's amazement.   □   In 1619, Descartes went through a "mental crisis." He had three dreams in which, according to his own interpretations, the Spirit of Truth showed him that he should set up a system of thought to direct mankind. He set up just such a system based on deduction, each known truth developing into another truth.   □   One of the incidents that really gave him the impetus to develop a new system of philosophy occurred around 1626 after he had just come to Paris. He was invited to hear a new system of philosophy discussed at the home of a cardinal. Descartes was unimpressed by this "new philosophy," said so, and proceeded to explain why. He impressed everyone so much that they all told him he should put his speech in writing.   □   From that time on, Descartes became a sort of philosophical hermit. He had a hard time writing around people—they tended to distract rather than help him—so he had his mail sent to one friend, had his money managed by another, and he went to Holland "for his health" (a statement that may have a good deal to do with the rather unhealthy situations of those who disagreed with the church).   □   Descartes spent the rest of his life publishing, hobnobbing with figures like Balzac, and arguing with every philosopher he could find. He died in Sweden in 1650 after contracting pneumonia on a 5:00 a.m. tutoring visit to the queen.

4-3

Motion with Constant Speed

Galileo tells us that velocity (speed) and changes in velocity, rather than natural position or place, are fundamental in describing motion. So we must now establish what we mean by **velocity**. If the velocity or speed of an automobile, for example, is constant, then we know that in equal times the car traverses equal distances. The velocity v is defined as the distance s divided by the corresponding time interval t:

$$v = \frac{s}{t} \qquad\qquad (4\text{-}1)$$

In physics, the conventional use of the word velocity is even more subtle than this. We can distinguish a velocity directed, say, due north from the same amount (same magnitude) of velocity directed east. In other words, at any given time the velocity of an object has both a **magnitude**, the numerical amount given by Equation (4-1), and a **direction**. We call such a quantity a **vector**. (The simple properties of vectors are discussed in Appendix 4A, pages 297–99.) If we want to work only with the magnitude of the velocity and disregard the direction, we call this the **speed** of the object. For now, to keep things as simple as possible, we will work with speed.

Since we want to be quantitative in our description of motion, we need some system of **units** that we can agree on to measure distance and time. In automobiles, the unit for distance is usually miles and, for time, hours, so that speed has the units miles per hour. Galileo might have measured speed in "cubits per pulse beat." However these are not useful units for most physics experiments since we would find it rather cumbersome to use them to express laboratory distances (lengths of tables, etc.) and times. The so-called "English" system of units uses the foot as the standard of length and the second for time, but is not so widely used in science as the mks (meter and second) or the cgs (centimeter and second) systems.

Example 4-1:

We can see from Equation (4-1) that if a car moves at constant speed for a distance of 10 meters (abbreviated m) in a time of 2 seconds (sec), then its speed is

$v = 10 \text{ m}/2 \text{ sec} = $ **5 m/sec**

Equation (4-1) is a relation between three measurable quantities: speed, distance, and time, and could be solved for any one of these. For example, by multiplying Equation (4-1) by t on both sides, we find an equation for the distance traveled in time t:

$s = vt$ (4-2)

Example 4-2:

An airplane travels at a constant speed of 1000 miles per hour (mi/hr) for 30 min. How far has it gone? Using $s = vt$ we find,

$s = 1000 \text{ mi/hr} \times \frac{1}{2} \text{ hr} = $ **500 mi**

(For a graphic representation of Equation (4-2) and Example 4-2, see Appendix 4B, pages 299–300.)

As a final example of the use of the constant velocity formula in Equation (4-1), we could solve for the time t, if we are given both the distance traveled and the speed of the motion. This would be obtained from Equation (4-2) by dividing each side of the equation by v:

$$t = \frac{s}{v} \qquad \qquad \textbf{(4-3)}$$

Example 4-3:

A billiard ball rolls at a constant speed of 2 ft/sec along a frictionless pool table 8 ft long. How much time does the ball take to go from one end of the table to the other? We use $t = s/v$ and obtain

$$t = \frac{8 \text{ ft}}{2 \text{ ft/sec}} = \textbf{4 sec}$$

In any problem involving *constant* velocity, two of the three quantities v, s, and t must be known. Then we can solve for the third by using Equation (4-1), (4-2), or (4-3), respectively.

An example of nearly constant velocity is the rolling of a smooth ball across an almost **frictionless** surface like ice, which offers no perceivable resistance to the motion. The air track shown in Figure 4-1 is a common laboratory demonstration.

A reverse-operated vacuum cleaner forces air inside a hollow metal beam which will form the air track itself. A series of small holes drilled in the beam permits the air to escape. If a glider just made to fit over one corner of the beam is placed on the air track, it will ride on a cushion of air which will be virtually frictionless. By leveling the beam, the glider, when moved, will traverse equal distances in equal times as seen in Figure 4-2.

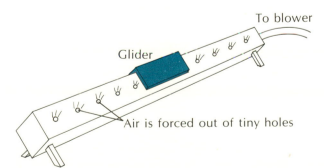

4-1 The air track is a nearly frictionless device used to demonstrate the principles of motion in the absence of resistance.

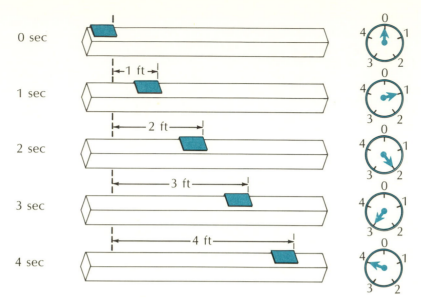

4-2 Motion at constant velocity. The glider is moving to the right at 1 ft/sec. Thus it traverses equal distances (1 ft) in equal times (1 sec).

4-4

Motion with Constant Acceleration

The **acceleration** of a body is defined as the change in velocity with time:

$$a = \frac{v}{t} \tag{4-4}$$

This equation means that for *constant* acceleration, the velocity must change by equal increments with equal time intervals. If a ball starts rolling from rest under constant acceleration and has a velocity of 1 ft/sec after an interval of 2 sec, it must be traveling 1 ft/sec *faster* or a total of 2 ft/sec after twice the time interval— after 4 sec. After 6 sec it will have speeded up to 3 ft/sec, and so on. The units of acceleration are distance/time/time or distance/(time)^2. So in the English system they are ft/sec^2; in the mks, m/sec^2; and in the cgs, cm/sec^2.

Example 4-4:

An automobile accelerated from rest to 60 mi/hr at a constant acceleration in 10 sec. What is the value of the acceleration? Use

$$a = \frac{v}{t} = \frac{60 \text{ mi/hr}}{10 \text{ sec}} = 6.0 \text{ mi/hr-sec}$$

Now the unit mi/hr-sec is rather cumbersome since it uses two *different* time units. It would be more consistent to change either $t = 10$ sec into hrs or to change $v = 60$ mi/hr into ft/sec, m/sec, or cm/sec. For example, if 60 mi/hr $= 88$ ft/sec,* then

$$a = \frac{v}{t} = \frac{88 \text{ ft/sec}}{10 \text{ sec}} = \mathbf{8.8 \text{ ft/sec}^2}$$

*See conversion unit table in Appendix 4C, page 301.

By multiplying both sides of Equation (4-4) by t, we arrive at an expression for the velocity under constant acceleration:

$$v = at \tag{4-5}$$

Example 4-5:

An earth-moon rocket accelerates at a constant acceleration of 1.5 m/sec^2. What is its velocity 60 sec after liftoff? Using Equation (4-5), velocity $v = at$ or

$$v = 1.5 \text{ m/sec}^2 \times 60 \text{ sec} = \mathbf{90 \text{ m/sec}}$$

In Example 4-5, the object to be accelerated started from a resting position. However, this is not always the case; acceleration can begin *after* an object already has some velocity, called the **initial velocity** $= v_0$. The increments of velocity are then merely added to (or subtracted from) v_0 to find the total velocity:

$$v = v_0 + at \tag{4-6}$$

Example 4-6:

An automobile is traveling at 30 ft/sec and starts to accelerate at a constant acceleration of 5 ft/sec^2. After 10 sec, what is its velocity? Using $v = v_0 + at$ with $v_0 = 30$ ft/sec, $a = 5$ ft/sec^2, and $t = 10$ sec, we find

$$v = 30 \text{ ft/sec} + 5 \text{ ft/sec}^2 \times 10 \text{ sec}$$
$$= (30 + 50) \text{ ft/sec} = \mathbf{80 \text{ ft/sec}}$$

Example 4-7:

An automobile traveling at 60 ft/sec is braked with a constant **deceleration** of 15 ft/sec^2. How long does it take to come to a stop? In this problem, we use Equation (4-6) with the *final* velocity $v = 0$ (since the automobile has come to rest) and with the *initial* velocity $v_0 = 60$ ft/sec. The acceleration 15 ft/sec^2 is in the *opposite* direction to the initial velocity v_0 and must therefore have the opposite sign from v_0. If we let

$v_0 = +60$ ft/sec, then we must set $a = -15$ ft/sec^2. Then using $v = v_0 + at$ or $0 = 60$ ft/sec $+ (-15)$ ft/sec$^2 \times t$ or $15 \times t = 60$,

$$t = \frac{60 \text{ ft/sec}}{15 \text{ ft/sec}^2} = \frac{4}{1/\text{sec}} = \frac{4}{1} \text{ sec} = \textbf{4 sec}$$

Equations (4-4), (4-5), and (4-6) related the physical quantities a, v, v_0, and t for constant acceleration. However, we very often cannot measure v or v_0 (except in automobile problems where we luckily have a speedometer reading). It is more common and much easier to measure the distance s rather than v. How is the distance s related to the four quantities a, v, v_0, and t? The derivation requires a few simple algebraic steps which are given in Appendix 4D (pages 301–302). There we find

$$s = v_0 t + \tfrac{1}{2}at^2 \tag{4-7}$$

Example 4-8:

The automobile in Example 4-7 decelerated from 60 ft/sec to rest in 4 sec. How far did it travel after deceleration began? With $v_0 = 60$ ft/sec, $a = -15$ ft/sec^2, and $t = 4$ sec,

$$s = (60 \text{ ft/sec} \times 4 \text{ sec}) + \tfrac{1}{2} \times (-15) \text{ ft/sec}^2 \times (4 \text{ sec})^2$$

$$= 240 - \frac{15 \times 16}{2} = 240 - 120 = \textbf{120 ft}$$

Although this is a bit more complicated than Equation (4-2) for distance at constant velocity, as in the former case, we can solve Equation (4-7) for any of the other quantities v_0, a, or t in terms of the three remaining ones. [Note that the equation for the time t gives a **quadratic** (squared) relation.]

It should be noted here that acceleration, like velocity, is also a vector quantity since both a magnitude and a direction are associated with it. In most of our discussion, we will deal only with the magnitude of the acceleration.

4-5

Galileo's Inclined Plane

To demonstrate that the motion of heavy objects downward due to gravity could be explained by the preceding formulas, Galileo described the following experiment in *Two New Sciences*:

> *A piece of wooden moulding or scantling, about 12 cubits long, half a cubit wide, and three finger-breadths thick, was taken; on its edge was cut a channel a little more than one finger in breadth; having made this groove very straight, smooth, and polished, and having*

lined it with parchment, also as smooth and polished as possible, we rolled along it a hard, smooth, and very round bronze ball. Having placed this board in a sloping position, by lifting one end some one or two cubits above the other, we rolled the ball, as I was just saying, along the channel, noting, in a manner presently to be described, the time required to make the descent. We repeated this experiment more than once in order to measure the time with an accuracy such that the deviation between two observations never exceeded one-tenth of a pulse-beat. Having performed this operation and having assured ourselves of its reliability, we now rolled the ball only one-quarter the length of the channel; and having measured the time of its descent, we found it precisely one-half of the former. Next we tried other distances, comparing the time for the whole length with that for the half, or with that for two-thirds, or three-fourths, or indeed for any fraction; in such experiments, repeated a full hundred times, we always found that the spaces traversed were to each other as the squares of the times, and this was true for all inclinations of the plane, i.e., of the channel, along which we rolled the ball. We also observed that the times of descent, for various inclinations of the plane, bore to one another precisely that ratio which, as we shall see later, the Author had predicted and demonstrated for them.

For the measurement of time, we employed a large vessel of water placed in an elevated position; to the bottom of this vessel was soldered a pipe of small diameter giving a thin jet of water, which we collected in a small glass during the time of each descent, whether for the whole length of the channel or for a part of its length; the water thus collected was weighed, after each descent, on a very accurate balance; the differences and ratios of these weights gave us the differences and ratios of the times, and this with such accuracy that although the operation was repeated many, many times, there was no appreciable discrepancy in the results.[2]

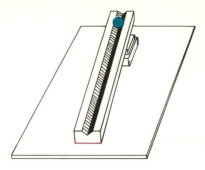

4-3 Galileo's inclined plane.

By this experiment, Galileo showed that the distance s traversed by the ball down the inclined plane was proportional to the square of the time t. Thus, motions of objects down the plane will always obey Equation (4-7) with the initial velocity v_0 equal to zero, or

$$s = \tfrac{1}{2}at^2 \qquad\qquad\qquad\qquad (4\text{-}8)$$

(See Appendix 4B, pages 299–300, for a graphic representation of this formula.)

But Equation (4-8) holds only for the special case of **constant acceleration**. Thus Galileo concluded that motion down an inclined plane is the motion of constant acceleration. Next, he raised one end of the plane and found that the same relation, s proportional to t^2, held. As he raised the plane more and more until it

[2]Arthur Beiser, *The World of Physics* (New York: McGraw-Hill, 1960), pp. 21–22. Used by permission.

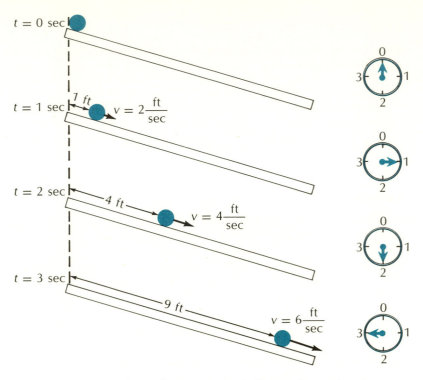

$t = 0$ sec

$t = 1$ sec   1 ft $v = 2\dfrac{\text{ft}}{\text{sec}}$

$t = 2$ sec   4 ft $v = 4\dfrac{\text{ft}}{\text{sec}}$

$t = 3$ sec   9 ft $v = 6\dfrac{\text{ft}}{\text{sec}}$

4-4 Uniformly accelerated motion. The ball goes down the plane at an acceleration of 2 ft/sec². After the first second, it has increased its velocity from zero to 2 ft/sec and has traversed 1 ft. After the next second, it is traveling 2 ft/sec faster or a total of 4 ft/sec and is 4 ft down the inclined plane. After the third second, it is moving at 6 ft/sec, which is 2 ft/sec faster than the previous value, and is 9 ft down the plane.

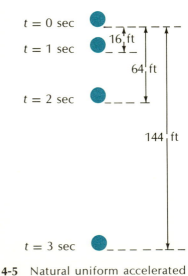

$t = 0$ sec

$t = 1$ sec   16 ft

$t = 2$ sec   64 ft

   144 ft

$t = 3$ sec

4-5 Natural uniform accelerated motion due to the gravitational pull of the earth.

was nearly vertical, he abstracted what the motion *would* be for objects falling vertically—*constant acceleration*. He called that value of the acceleration for pure vertical motion the **acceleration of gravity**.

A sequential diagram of the constant acceleration of an object is shown in Figure 4-4. When $a = 2$ ft/sec², $s = \frac{1}{2}at^2$ becomes numerically $s = t^2$. Note that the velocity is also shown in the figure, and that the object receives equal increments of velocity (increments of 2 ft/sec) in each unit of time (1 sec).

For freely-falling objects, the acceleration of gravity is given the special symbol g and its value on the surface of the earth is $g = 32$ ft/sec² $= 980$ cm/sec² $= 9.8$ m/sec². Thus, according to Equation (4-8), in the first second an object dropped from rest will fall 16 ft; at the end of the next second it will have traveled a total of $4 \times 16 = 48$ ft; after the third second it will be $9 \times 16 = 144$ ft from the starting point (see Figure 4-5).

Example 4-9:

How far will the object discussed above fall in 5 sec? Using $s = \frac{1}{2}at^2$, with $a = g = 32$ ft/sec^2,

$s = \frac{1}{2} \times 32$ ft/sec$^2 \times (5$ sec$)^2$

$= 16 \times 25 =$ **400 ft**

We can also use Equation (4-7) if the object is given an initial velocity upward or downward.

Example 4-10:

A ball is thrown straight upward with an initial velocity of 4.9 m/sec. How far will it travel in 1 sec? Use $s = v_0 t + \frac{1}{2}at^2$ with $v_0 = 4.9$ m/sec, $t = 1$ sec, and $a = g = -9.8$ m/sec^2. The minus sign appears here since the acceleration downward (due to gravity) is oppositely directed from the initial velocity (upward). As a consequence, positive values of s mean the ball is above the starting point, while negative values mean it is below the starting point. Then

$s = 4.9$ m/sec $\times 1$ sec $- \frac{1}{2} \times 9.8$ m/sec$^2 \times (1$ sec$)^2$

$= 4.9$ m $- 4.9$ m $=$ **0 m**

But how can this result be true? An answer of zero means the ball is at the original starting point which could only be true if the ball had gone up and come back—just as it passed the original point from which it was thrown. In Example 4-10, at any time t less than 1 sec, the ball would be above the starting point, and for any t longer than 1 sec, it would be below. Let's see if this is true in another example:

Example 4-11:

Solve the previous problem for $t = \frac{1}{2}$ sec. Then

$s = 4.9$ m/sec $\times \frac{1}{2}$ sec $- \frac{1}{2} \times 9.8$ m/sec$^2 \times (\frac{1}{2}$ sec$)^2$

$= 2.450$ m $- 1.225$ m $=$ **1.225 m**

and we see that the ball is *above* the starting point (see Figure 4-6).

The reader should prove to himself that for larger times (say $t = 2$ sec), the ball is below the starting point (assuming it has not hit the ground first).

Example 4-12:

You may have thought of the question: At $t = \frac{1}{2}$ sec, is the ball at the highest point of its trajectory? The answer (yes) can be deduced from

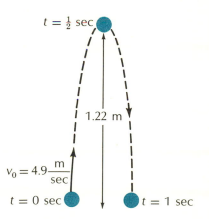

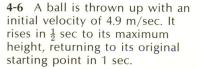

4-6 A ball is thrown up with an initial velocity of 4.9 m/sec. It rises in $\frac{1}{2}$ sec to its maximum height, returning to its original starting point in 1 sec.

the symmetry of the upward and downward motion. On the way down the ball just repeats its upward motion (like a movie run backwards).

We can also see this from Equation (4-6) and Example 4-10. The final velocity at the top of the trajectory is $v = 0$. So we see

$$0 = v_0 + at = 4.9 \text{ m/sec} - 9.8 \text{ m/sec}^2 \times t$$

Solving for t,

$$t = \frac{4.9}{9.8} \text{ sec} = \tfrac{1}{2} \text{ sec}$$

4-6

Compound Motion

One of the important practical problems in Galileo's time was the calculation of the **trajectories** of cannonballs fired from artillery pieces. Many mathematicians had worked on this problem, but they were unable to find the correct solution until Galileo discovered the laws of motion for terrestrial objects. A cannonball, or any object that is fired or thrown, will be pulled down toward the earth while traveling on its trajectory. Galileo concluded that, except for the unknown effects of air resistance, the trajectory of an object could be explicitly calculated and its point of impact accurately determined if the object's vertical motion downward was considered to be independent of the object's horizontal motion. Copernicus had already alluded to the fact that the circular motion of the earth (i.e., the horizontal motion of an object *on* the earth) was independent of upward or downward motions. Galileo refined this idea and obtained quantitative results that could be calculated mathematically.

He had already concluded that pure horizontal motion is described in terms of constant velocity motion [Equations (4-1), (4-2), and (4-3)], while his experiments with the inclined plane had shown that pure vertical motion is constant accelerated motion [Equations (4-4)–(4-8)]. Thus he concluded that any complicated projectile motion on or near the earth's surface could be determined by using $s_{\text{horiz}} = vt$ [from Equation (4-2)] for the horizontal distance traveled and $s_{\text{vert}} = v_0 t + \tfrac{1}{2}gt^2$ [from Equation (4-7)] for the vertical distance. Since we are dealing with constant acceleration due to the earth's gravity, from now on we will use the symbol g for the acceleration a.

Example 4-13:

A cannon with a muzzle velocity of 500 ft/sec fires a shell aimed horizontally from the top of a cliff 64 ft above level ground. How far from the base of the cliff will the shell hit the ground?

The vertical and horizontal motions of the shell are separate (see

Figure 4-7); we will consider the vertical one first. We solve for time t, the only common quantity between the vertical and horizontal motions. The initial *vertical* velocity is zero, so we modify Equation (4-8): $s_{vert} = \frac{1}{2}at^2 = \frac{1}{2}gt^2$. If $s_{vert} = 64$ ft and $g = 32$ ft/sec^2,

$$64 \text{ ft} = \frac{1}{2} \times 32 \text{ ft/sec}^2 \times t^2$$

Solving for t^2, we obtain

$$t^2 = \frac{2 \times 64}{32} = 4 \text{ sec}^2 \text{ or } t = 2 \text{ sec}$$

Now if the shell takes 2 sec to fall to the ground, we can use this result in the formula for constant velocity to obtain the *horizontal* distance it traveled before hitting the ground. (Remember that the horizontal velocity = v_{horiz} = 500 ft/sec.)

$$s_{horiz} = v_{horiz} \times t = 500 \text{ ft/sec} \times 2 \text{ sec}$$

$$= \textbf{1000 ft from the base of the cliff}$$

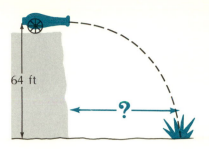

4-7 A cannon is fired horizontally from a cliff 64 ft above ground level. Where does the projectile hit the ground?

More complicated trajectories can be calculated from the basic equations (4-1)–(4-7). For instance, if the vertical motion has an initial velocity upward, Equation (4-7) would be used to find the vertical motion and Equation (4-2) to find the horizontal motion.

A lecture demonstration sometimes called "the monkey and the hunter" is often used to illustrate compound motion. A hunter aims his gun at a monkey sitting on a tree branch. Realizing his predicament, the monkey waits until he sees the flash from the muzzle of the hunter's gun and then drops from the branch to the ground, hoping that he will not be hit by the bullet which he believes will pass over his head. Does the bullet hit the monkey?

Yes, but because the hunter has poor aim, the bullet only severs the monkey's banana in half, missing the monkey completely. The monkey accelerates downward at the same rate (the same distance in the same time) as the bullet accelerates *down* from its original line-of-sight trajectory. By the time the bullet reaches a point directly under the monkey's original position on the tree branch, the bullet will have accelerated down from the original line of sight for the *same* amount of time (and thus for the same vertical distance) as the monkey has fallen. So they must meet, as shown in Figure 4-8. After the bullet is fired, the monkey lets go of the branch and starts to fall (Figure 4-8 b). After some time interval (say, $\frac{1}{10}$ sec), assume the bullet is halfway to the monkey. According to Equation (4-8) and Example 4-13, the bullet will have accelerated down from its original line-of-sight path by

4-8a A rifle is aimed at the monkey.

$$s = \frac{1}{2}gt^2 = \frac{1}{2} \times 32 \text{ ft/sec}^2 \times (\tfrac{1}{10} \text{ sec})^2 = 0.16 \text{ ft} = 1.92 \text{ in.}$$

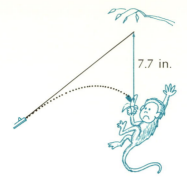

4-8b As the bullet leaves the gun barrel, the monkey starts to drop from the tree.

4-8c After $\frac{1}{10}$ sec the monkey has dropped 1.9 in. from the branch and the bullet has fallen the same amount from its original line-of-sight path.

4-8d After $\frac{2}{10}$ sec, the monkey has dropped 7.7 in. and the bullet has fallen the same amount from its original path.

(Remember, g = 32 ft/sec² on the earth's surface.) But the monkey has also dropped the same distance (Figure 4-8 c). In this example, when the bullet is just below the tree branch, it is $2 \times \frac{1}{10} = \frac{1}{5}$ sec after firing. The bullet has now dropped

$$s = \tfrac{1}{2}gt^2 = \tfrac{1}{2} \times 16 \text{ ft/sec}^2 \times (\tfrac{1}{5} \text{ sec})^2 = 7.68 \text{ in.}$$

from its original line-of-sight path. But *so has the monkey* and the two meet, and the monkey loses the top part of his banana.

4-7

Galileo and the Inertia Principle

Inertia is a property of any object. It involves both the amount of motion (velocity) and the matter content (what we now call **mass**) of an object. Although Newton was the first to deal properly with the concept of mass itself, Galileo and Descartes both made important contributions to the early work on inertia. The so-called "law of inertia" is simply the statement that an object resists any change in its state of motion (or state of rest if it is not moving). The degree of resistance an object offers is directly proportional to the amount of ponderable matter it contains. The matter content of an object, its mass, will be discussed later in Chapter 6. For now, we can see that, in parts, Galileo's earlier quotation ". . . along a horizontal plane the motion is uniform" and "would carry the body at a uniform rate to infinity" (page 38) is a statement of the principle of inertia.

Actually, Galileo believed in a circular kind of inertia along the earth's surface for terrestrial objects. He was following Copernicus in this idea. As we saw in Chapter 3, Copernicus suggested that the circular motion of the earth (and the objects on the earth) was a natural motion which would continue indefinitely. This was the beginning of the idea which Galileo extended to any motion of any object, not merely the motion resulting from the earth's rotation.

René Descartes formulated the problem in a different way, coming even closer to what we now consider to be the correct law of inertia. In Principle 37 of his *Principles of Philosophy*, he stated:

> *The first law of nature: That each thing stays in the state in which it is, so long as nothing changes it. . . .* [*And he goes on to explain,*] *From the fact that God is not subject to change, and that He acts always in the same way . . . each thing continues to be in the same state insofar as it is able . . . and that if it is at rest it does not start to move by itself. But when it has once started to move . . . it continues afterwards to move and it never comes to rest by itself.*

Notice that instead of using Galileo's experimental approach, Descartes argued deductively from what he believed to be the first principle—the immutability of God. Descartes also arrived at the correct concept of **linear inertia**, which had eluded Galileo, in the same manner:

> *The second law of nature: That every body which moves, tends to continue its movement in a straight line.*

However, in his later Principles concerning the collision of objects, Descartes ran into trouble employing deductive techniques. Newton eventually formulated the problem correctly by combining both of Descartes' laws of nature into his own first law of motion, as we will see in Chapter 6.

Another aspect of the inertia problem concerns the falling rates of two objects of different weights. As we saw in Chapter 2, Aristotle would have had us believe that a heavier object falls faster than a lighter one and in a time inversely proportional to its weight. Galileo argued against this in the following passage from *Two New Sciences*:

> *Aristotle says that "an iron ball of one hundred pounds falling from a height of one hundred cubits reaches the ground before a one-pound ball has fallen a single cubit." I say that they arrive at the same time. You find, on making the experiment, that the larger outstrips the smaller by two finger-breadths, that is, when the larger has reached the ground, the other is short of it by two finger breadths; now you would not hide behind these two fingers the ninety-nine cubits of Aristotle, nor would you mention my small error and at the same time pass over in silence his very large one. Aristotle declares that bodies of different weights, in the same medium, travel (insofar as their motion depends upon gravity) with speeds which are proportional to their weights; this he illustrates by [the] use of bodies in which it is possible to perceive the pure and unadulterated effect of gravity, eliminating other . . . influences which are greatly*

dependent upon the medium which modifies the single effect of gravity alone. Thus we observe that gold, the densest of all substances, when beaten out into a very thin leaf, goes floating through the air; the same thing happens with stone when ground into a very fine powder. But if you wish to maintain the general proposition you will have to show that the same ratio of speeds is preserved in the case of all heavy bodies, and that a stone of twenty pounds moves ten times as rapidly as one of two; but I claim that this is false and that, if they fall from a height of fifty or a hundred cubits, they will reach the earth at the same moment.[3]

It is said that Galileo proved that two objects of different weights reach the ground at the same time when he dropped them off the Leaning Tower of Pisa. This is probably historical fiction, perhaps perpetrated by one of Galileo's admiring students or biographers. In any event, Galileo did realize that the content of matter—the weight of an object—should not affect the time of its descent in a gravitational free-fall: ". . . in a medium totally devoid of resistance, all bodies would fall with the same speed." The changes of motion due to gravity are the same for all objects. Another way of saying the same thing is that the acceleration of gravity is the same for all bodies near the earth's surface. A deeper understanding of the reasons for this could not be gained until Newton had provided a definition for mass or the content of matter.

4-8

Galileo's Difficulties

A few historical notes on Galileo and the events and people of his time are important here. It was very difficult for Galileo and Descartes, who were each independently looking for a solution to the problem of accelerated motion, to divorce themselves from the notion that place or distance was the relevant quantity. They each "proved" that constant acceleration (which, as we know, is equal increments of velocity in equal times) meant "velocities proportional to distance traversed" and they each further "proved" that this led directly to distance proportional to (time)^2. Using a little bit of algebra and Equations (4-5) and (4-8), it is clear that velocity is actually proportional to the *square root* of the distance. Descartes never realized his errors—one in logic and one in simple mathematics. In later years, Galileo did realize his error and corrected it when he published *Two New Sciences* in 1638. The common error both made was in beginning from "velocity proportional to distance traversed," rather than "velocity proportional to time elapsed."

[3]Arthur Beiser, *The World of Physics* (New York: McGraw-Hill, 1960), p. 12. Used by permission.

The idea of time as a relevant physical quantity was too novel for these men to grasp; time had not been commonly thought of in this way before. But Leonardo da Vinci had known this, writing more than a century before Galileo:

> *The heavy body which descends at each degree of time acquires a degree of movement [velocity] more than in the degree of time preceding, and similarly a degree of swiftness greater than the degree of the preceding movement. . . . Let us say that in the first degree of time it acquires a degree of speed, in the second degree of time it will acquire two degrees of speed, and so it continues in succession.*

Constant rechecking of his work undoubtedly eventually enabled Galileo to see his error.

Galileo's problems with the Catholic Church and the Inquisition stemmed from his enthusiasm for the Copernican theory, which he was teaching and publishing, and the observations he made with his newly developed telescope in 1609. Several of his observations, particularly those that showed moons circling the planet Jupiter and the phases of Venus, were recorded in *The Starry Messenger* and published in 1610. These results were subsequently to be verified by other observers and were to virtually demolish the church's arguments (particularly those of influential Aristotelians—the professors at the universities) against the Copernican system.

Galileo was first warned by the Inquisition in 1616 when *De Revolutionibus* was placed on the index of banned books. (It was not removed until 1838!) His writings, especially *Dialogue on the Two Chief World Systems* (1632), although at first approved by the church, were later to lead to his trial and sentence by the Inquisition.

In 1633, the 70-year-old Galileo was forced to kneel before the Cardinals and Inquisitors and to recant the heresies he had promoted in teaching the Copernican system:

> *I have been enjoined by this Holy Office altogether to abandon the false opinion which maintains that the sun is the center and immovable . . . and that the earth is not the center, and movable. . . . I abjure, curse, and detest the said errors and heresies . . . and I swear that I will never more in the future say, or assert anything, verbally or in writing, which may give rise to a similar suspicion of me.*[4]

Bruno was burned at the stake; Galileo was imprisoned. Who was next? Kepler, as we will see, dodged the Inquisition by obtaining protection in various Protestant countries, and Descartes (also a Copernican, but living in relative safety in Holland) ap-

[4]American Foundation for Continuing Education, *Exploring the Universe* (New York: McGraw-Hill, 1963), p. 135. Used by permission.

pended to the end of his work on physics in the *Principles of Philosophy:*

> *At the same time, recalling my insignificance, I affirm nothing, but submit all these opinions to the authority of the Catholic Church, and to the judgment of the more sage; and I wish no one to believe anything of what I have written, unless he is personally persuaded by the force and evidence of reason.*

These were indeed hard times for physicists to be free with their ideas.

Questions

1 Galileo helped immensely in defining physics (i.e., giving a much clearer picture of what things were properly the domain of physics). Name some of the conceptual devices he introduced and explain how they represented an improvement over previous ideas.

2 Galileo revolutionized astronomy. What were some of his contributions?

3 We can demonstrate the principle that gravitational acceleration is independent of mass by dropping a 6-in. and a 3-in. cannonball simultaneously from the balcony of the Leaning Tower of Pisa. But if we drop a feather and a cannonball, the cannonball will always fall faster. Does this experiment disprove our principle? Explain.

4 What do you think constitutes a valid relationship between an experiment and a theory. Try to explain why the feather-cannonball experiment in Question 3 fails *or* why the cannonball-cannonball experiment is successful.

5 How would Aristotle have explained the constant velocity of the glider moving along the air track (Figure 4-1) if he had seen this demonstration in his time?

6 Galileo used the following argument against Aristotle's theory of falling bodies: If we have two stones, a heavy one A and a light one B, then A, being heavier, will fall faster than B. If we tie A and B together, B, which falls more slowly, will retard A and the combination $A + B$ will fall more slowly than A alone. Similarly the combination $A + B$ will fall more rapidly than B alone. But A and B together make a heavier object than A alone, and so, according to Aristotle's theory, $A + B$ should fall faster than A. Does the situation described successfully disprove Aristotle's theory or was Galileo also mistaken? Does the theory invite contradiction? Explain.

7 If a body moving along a straight line reverses the direction of its motion twice, has the acceleration of the object changed or could it have been constant? Explain.

8 Can we account for the effect of air resistance on falling objects by simply reducing the numerical value of the acceleration due to gravity? Why or why not?

9   If the acceleration of an object has constant magnitude and direction, will its path necessarily be a straight line? Give an example to support your answer.

10 If an object travels in a straight line, what can we conclude about the relationship between its velocity and its acceleration?

11 According to Ptolemy, all of the other planets revolved around the earth. How did Galileo's observation that moons circled around Jupiter while it traveled in its orbit deal a crushing blow to the Ptolemaic arguments for an earth-centered universe?

12 Explain what is meant by abstraction in terms of Galileo's new way of looking at motion.

13 Do you think that Galileo's inductive method based on experiment or Descartes' deductive method based in irrefutable premises eventually led to a closer knowledge of the ultimate truth about the universe. Explain.

Questions Requiring Outside Reading

1   How was Galileo able to defend his Copernican views prior to his trial in 1633?

2   How did Descartes view Galileo, and vice versa? Did they know of one another's work on falling bodies?

3   The so-called "fundamental" properties of the physical world are mass, length, and time. Can you explain why these particular properties of objects and events were chosen? [An excellent discussion of perception in physics can be found in the Appendix to *The Special Theory of Relativity*, by David Bohm (New York: Benjamin, 1965) nontechnical.]

4   What other observations did Galileo discuss in *The Starry Messenger* in addition to Jupiter's moons? How did his writings affect Aristotelian cosmology?

5   What other contributions did Galileo make to science besides the ones mentioned in this chapter?

6   What contributions did Descartes make to science and to mathematics?

Problems

1   An interplanetary rocket ship travels with constant velocity, covering a distance of 5×10^8 mi in 200 hrs. What is its speed?

2   Chitty-Chitty-Bang-Bang, World Land Speed Record holder several years ago, could travel at 120 mi/hr. How long did the car take to cover the flying mile?

3   John lives a block (400 ft) from the bus stop. If the bus leaves precisely at 7:42 and John walks at 5 ft/sec, what is the latest time he can leave home and not have to run?

4   Throckmorton P. Twobody, working high on a TV broadcasting tower, drops a wrench. Using the kinematic formulas for motion with

constant acceleration, find (a) how fast the wrench is traveling when it passes 6 in. in front of the boss' nose 200 m below Throckmorton's hand and (b) how long it took the wrench to fall the 200 m.

5   A ship covers 550 nautical mi in one day and 9 hrs. What is its average speed in knots if 1 knot = 1 nautical mi/hr?

6   A car with an initial speed of 20 ft/sec accelerates uniformly until its speed is 80 ft/sec. With an acceleration of 5 ft/sec, how long must the car accelerate to reach this speed?

7   What distance does the car in Problem 6 cover during acceleration?

8   Throckmorton, on his way home after being fired (see Problem 4), is riding his motorcycle at 50 m/sec (that's fast!). Seeing a red light ahead, he applies the brakes and comes to a stop after decelerating uniformly for 50 m. (a) What was the value of the deceleration? (b) How long did the bike take to come to rest?

9   A baseball player throws a ball straight up in the air and catches it again at the same height he released it. If the ball spends 2.0 sec in the air, how high above his hands did it rise? (The acceleration of gravity is 9.8 m/sec².)

10 Greg shoots a marble from his slingshot *directly at* a tomcat on a fence. The cat and the slingshot are on the same level. The cat is 10 m from Greg and the initial (horizontal) velocity of the marble is 20 m/sec. Neglecting air resistance, find (a) how long it takes the marble to reach the fence and (b) how far below the cat the marble hits the fence (i.e., how far it has dropped below the point of Greg's original aim).

11 In his new job as a bartender, Throckmorton slides a mug of beer down the length of the bar to a customer. Unfortunately, he slides it too fast and it overshoots, landing on the floor 5 ft away from the edge of the bar. If the bar is 4 ft high, how fast was the mug traveling when it left the bar and started to fall?

12 Napoleon and the Duke of Wellington line their artillery up against one another on two hills of exactly the same height. Two of the cannons are aimed directly at each other and are fired simultaneously. Napoleon's cannon has a muzzle velocity of 1000 ft/sec; Wellington's is 500 ft/sec. The distance between the cannons is 6000 ft. Will the cannonballs hit their respective targets? If so, how far from and how far below the crest of each hill will they land?

13 Beginning from a rest position, a freight train accelerates uniformly from the railroad switchyard at a rate of 0.25 m/sec². A hobo initially 100 m behind the train runs to catch it. If he can run 4 m/sec, (a) can he catch the train and, if the answer to (a) was yes, (b) how fast will it be moving when he reaches it?

14 Prove that for constant acceleration the velocity is proportional to the *square root* of the distance traversed, and not to the distance itself as Descartes and young Galileo believed.

Kepler and the Discovery of Planetary Orbits

5

As we saw in Chapter 4, Galileo contributed the first empirical formulas or "laws" for terrestrial motion: constant velocity for horizontal motion and constant acceleration for vertical motion. These empirical formulas were to form the first part of the substructure of the mechanical revolution in physics.

The second part of this substructure was discovered by Johannes Kepler—a contemporary of Galileo although the two never met and only corresponded occasionally. Kepler's great contributions, like those of Galileo and Copernicus, required creative imagination coupled with exceedingly painstaking calculation. In the tradition of the Pythagoreans, Kepler was interested in finding natural "laws" or, as he put it, "harmonies" of planetary motion. He believed in the Copernican theory like Galileo, but felt it could be simplified if epicycles and the other constructs that were still required to describe oval-shaped orbits with perfect circles could be replaced—but replaced with what?

5-1

Introduction

Kepler needed extremely accurate astronomical data on the motion (positions and times of observation) of the various planets for his work. Luckily such data just happened to be available at the time. Tycho Brahe had the world's largest, best-equipped, and most accurate observatory. His data, particularly on Mars' orbit, were amazingly accurate in view of the fact that the invention of the telescope was still 20 years in the future. With the unaided eye, Brahe used huge wooden astronomical quadrants and other devices (see Figure 5-1) calibrated to great precision in fractions

5-2

Kepler and Tycho Brahe

Tycho Brahe by M. van Mierevelt.

The Royal Society, photograph John R. Freeman and Co., Ltd.

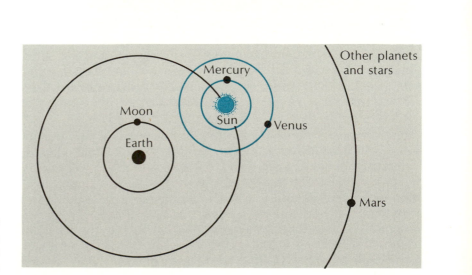

5-1 Tycho Brahe's picture of the solar system. Notice the earth is still stationary and in the center of the universe.

BRAHE

Tycho Brahe was one of the most unusual characters in the history of science. From the very beginning, his life was full of bizarre incidents. □ He was born in 1546, the first son of a noble Danish family. His uncle, George, wished to adopt him, an idea the family greeted with something less than enthusiasm. Undaunted, the stubborn uncle kidnapped Tycho and locked the infant in his castle. A family quarrel resulted, but when the Brahes had another son they decided to let uncle George have his way. □ Tycho first became interested in science in 1560 when an eclipse occurred according to astronomical predictions. He was so fascinated with the idea of foretelling the future that, against his family's advice, he began to study science. □ Tycho had a silver nose—the result of a duel with another nobleman on some point of mathematics. Apparently a better scientist than a swordsman, Tycho was forced to employ his metallurgic skill in replacing his nose. (He carried around a bottle of glue in case it fell off.) □ By 1576, Tycho had become moderately famous in Europe as an astronomer and decided to leave Denmark and settle in Germany. But King Frederick, who was very interested in the sciences, offered him a small island off Hamlet's Elsinore call Hven, a permanent salary, and an observatory built to whatever specifications he wished—an offer Brahe could hardly decline. □ Since Tycho had access to almost unlimited funds, the island became a place where he acted out his architectural fantasies from special rooms to imported birds for his imported trees. The selection of the island—usually a very cloudy and foggy one—was somewhat odd, but Tycho seems to have overcome this difficulty with grace. □ For the next 17 years Brahe worked with the instruments he had designed, observing the motions of the planets and the positions of the stars, and recording them all accurately. He drew up occasional horoscopes for the king's sons, all of which proved to be accurate even to the death of one of them at 18. □ Unfortunately, kings and their favors do not last forever. Frederick was replaced by King Christian, who was more interested in good, efficient government than in science. Brahe was accused of appropriating fields that were not his, firing persons, and sleeping with a concubine (the state of Brahe's marriage is somewhat in doubt). □ The crown decided to revoke his income and his other financial support. Dismayed, Brahe left Denmark and ended up in Poland where the king, more interested in science than in government, offered Brahe a position. There Brahe met Kepler, and they worked together until Brahe's death in 1601. □ Tycho was buried with the incredible pomp and ceremony befitting one of the hardest-working astronomers and greatest observers of all time. His silver nose, however, was stolen by graverobbers.

of a degree. Tycho had amassed a wealth of data on Mars' orbital motion, but was not enough of a mathematician to properly employ his findings. Even if he had been, he would not have made much headway for he did not believe in the Copernican theory.

Brahe had developed his own explanation of the planetary orbits (Figure 5-1). The advantage his system had over the Ptolemaic universe was that, if Mercury and Venus rotated around the sun (which, in turn, rotated about the stationary earth) as he claimed, neither planet would be very far from the sun, a fact in accord with actual observation. (In the Ptolemaic model, special epicycles were required for these two planets to be seen near the

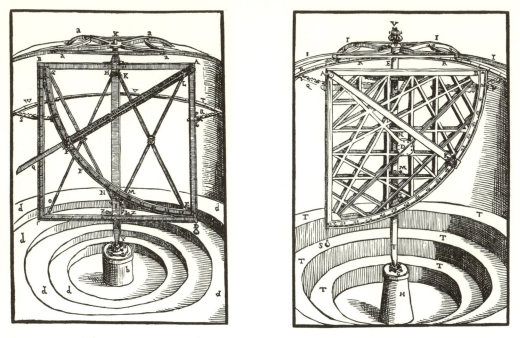

5-2 Some of the instruments Tycho Brahe used in his astronomical observations.

sun.) Venus and Mercury always appear as morning or evening stars; they are fairly close to the sun at all times but are seen only when it is dark enough, near sunrise or sunset.

Most of Brahe's data was obtained in Hven, a small island between what are now Sweden and Denmark. However, when King Frederick II died, Tycho lost his position as Imperial Mathematician and moved to Prague where, financed by Emperor Rudolph II, he built another observatory. To fit his data to his own model of the universe, Tycho needed the help of an expert mathematician. Kepler, on the other hand, wanted to see what regularities or harmonies he could find by fitting Tycho's invaluable data into the Copernican universe. So it is not too surprising that when young Kepler arrived at Tycho's observatory (the "celestial palace") in Prague on January 1, 1600, each in his own way hoped a fruitful collaboration was about to begin. Tycho guarded his data closely and jealously, however, and Kepler finally gained access to it only after Brahe's death a year later.

A long, tortuous series of trial-and-error calculations then began as Kepler wrestled with the Mars problem. Unlike any other famous physicist before or since, Kepler kept a written running account of the errors and successes he made and the blind alleys he reached in his work which is fascinating to read. During almost

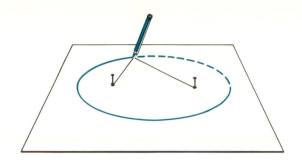

5-3 Drawing an ellipse. A string is tied loosely to two fixed pins or nails. A pencil, held so that the string is taut, will trace out an ellipse.

20 years of painstaking labor, Kepler arrived at three "laws" of planetary motion, which occupy an equal place beside Galileo's "laws" of terrestrial motion. These empirical formulas were essential to Newton; without them he could not have effected the synthesis that was to be the Mechanical Revolution.

First Kepler broke with the traditional thinking of Galileo, Copernicus, Brahe, and, of course, Ptolemy and Aristotle by announcing that the planets did not move in perfect circles! Instead he said the orbit of a planet was an **ellipse**, with the sun at one focus. An ellipse is the figure obtained when you tie a string between two fixed points and, with the string taut, draw the resulting curve (see Figure 5-3). With this one "law," the cobwebs of all the epicycles of the past were swept away. The single ellipse did all and even more than many, many epicycles ever could. It must have been as hard for Kepler to discard circular planetary motion as it was for Galileo to break with the description of terrestrial motion as being dependent on position. A particularly difficult problem in Kepler's situation was what to do with the other **focus** of the ellipse. If the sun is at one focus, what is at the other (Figure 5-4)? As Kepler finally realized, *nothing at all occupies the other focus!*

A planetary orbit with no astronomical body occupying one focus was not symmetrical and therefore difficult to accept. It certainly did not compliment Aristotle's principles. Mathematically each focus in the ellipse was of equal importance. Kepler had laboriously tried other noncircular orbits which, unlike the ellipse, had the virtue of requiring only one "center" or focus. In particular, he worked with the ovoid, an egg-shaped orbit, but found it so difficult to deal with that he repeatedly *approximated* its shape using the mathematically tractable *elliptic* figure. After many years he finally came to the realization that this approximation method—the ellipse—was what he was really looking for all the time! And so Kepler's first law was "discovered."

5-3

The Three Laws

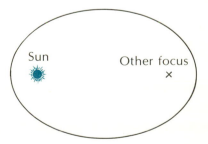

Sun   Other focus

5-4 Kepler's planetary elliptical orbit. There are two foci: one at the position of the sun, the other with no astronomical body.

Johannes Kepler

The Granger Collection

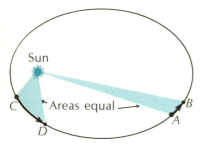

5-5 Kepler's second law: equal areas in equal times. The two colored regions have equal areas if the time for motion from *A* to *B* is the same as the time for motion from *C* to *D*.

The second "law" was also the result of long, hard calculations on the orbit of Mars. Kepler found that, in the elliptical orbit around the sun, a planet sweeps out equal areas in equal times. Thus if it takes one month to go from *A* to *B* and also a month to go from *C* to *D*, the areas swept out by the planetary orbit as measured from the sun (the colored regions in Figure 5-5) are equal.

Kepler's third law relates data on the orbit of one planet to the orbit of another planet. Using Tycho's dedicated observations and many of his own calculations, Kepler found that the ratio of the *square* of the time **period** for a complete orbit *T* to the *cube* of the planet's average orbital **radius** from the sun *R* is the same for all planets. That is, R^3/T^2 is constant for all planets in our solar system.

Kepler described his trial-and-error discovery method in the *New Astronomy* (1609) which contains his first two laws:

> *If thou dear reader art bored with this wearisome method of calculation, take pity on me who had to go through with at least seventy repetitions of it, at a very great loss of time; nor wilst thou be surprised that by now the fifth year is nearly past since I took on Mars.*

KEPLER   Johannes Kepler (1571–1630) had a rather unhappy family life in Germany in his early years. When he was only three, his father left home to fight in the Netherlands. He finally returned, broke and disgusted (having nearly been hanged for robbery), bought a beer-hall, and put Johannes to work as a busboy. □ After attending a monastery school, where he seems to have been somewhat disliked by the other students largely because of his desire to please the authorities, Kepler went to Graz to teach. One of his tasks there was to compile a yearly astrological calendar. Astrology at that time was still a well-respected science and this was an extremely important job. Much to his surprise, Kepler, in his first calendar, correctly predicted both the invasion of the Turks and a bitter cold spell, earning him a good deal of notoriety throughout Austria. □ His job at Graz came to an abrupt end when he and the other citizens of the city were ordered by the new archduke to leave within 45 days if they did not become Catholics. After a short period of financial crisis, Kepler went to work for Tycho Brahe in Prague. □ Biographers differ in their feelings about how the two got along. Kepler seemed to be a much more sensitive man than the irritable and erratic Tycho. After Brahe's death, Kepler had enormous difficulty with Brahe's family when he published their observations, especially since he used them to prove the Copernican theory in which Brahe himself had never believed. □ Meanwhile, Kepler went through six troubled years trying to prevent his mother from being executed as a witch. By the time she went to trial, Mrs. Kepler had been accused of causing the sickness and death of animals and people all over the countryside. She probably would have been tortured until she confessed and then burned at the stake, but Kepler, at the time the Imperial Mathematician, intervened and had the case transferred to calmer country where his mother was eventually freed. □ Kepler had financial problems all his life. He was constantly receiving promises of money from emperors and ministers who either died or were deposed before the payments were made to him. He was a hypochondriac, always sick or imagining that he was sick. He also apparently had an aversion to water, and reportedly never took a bath—perhaps not an unrelated malady. □ He was tremendously disorderly and once said that if anything orderly came out of him, it had been begun ten times. In actual fact, when he computed his tables for the orbit of Mars (which baffled astronomers and drove Rheticus, Copernicus' protégé, insane), he made the same calculations 181 times apiece for at least 40 computations.

Kepler repeatedly reached blind alleys in his calculations and often formulated new hypotheses based on thousands of hours of painstaking work that he was sure were correct, only to follow them in some succeeding chapter with a statement like

> Who would have thought it possible? This hypothesis, which so closely agrees with the observed oppositions [positions of Mars], is never the less false.[1]

In his last work, *Harmony of the Worlds* (1619), Kepler published his third law along with some other irrelevant and incorrect discoveries pertaining to planetary orbits. The discovery that Kepler believed to be his greatest achievement turned out to be

5-4

The Harmony of the Worlds

[1]Arthur Koestler, *The Watershed* (New York: Doubleday, 1960).

5-6 The five regular geometric solids.

Tetrahedran   Cube   Octahedron   Dodecahedran   Icosahedran

meaningless. There were in Kepler's time only six known planets: Mercury, Venus, Earth, Mars, Jupiter, and Saturn, in the order of their distance from the sun. Kepler knew that there were five (and only five) regular geometric solids. [By regular, we mean that each face is a regular polygon (all sides are of equal length, like an equilateral triangle or a square). These solids (see Figure 5-6) are the tetrahedron (4 sides), cube (6 sides), octahedron (8 sides), dodecahedron (12 sides), and the icosahedron (20 sides).]

Kepler believed he had discovered the fundamental secret of the universe when he found that if the orbit (thought of as a sphere) of the innermost planet, Mercury, were *inscribed* inside the octahedron (the orbit being tangent to all the plane faces), then the orbit of Venus would be *circumscribed* about the same solid (its orbit just touching the outer corners). Venus' orbit would then be inscribed inside the dodecahedron and the earth's orbit circumscribed about it, and so forth until all of the five figures were used to determine the orbits of the six known planets.

Unfortunately this system does not work very well in light of the modern data on planetary distances, and it breaks down completely if we include the asteroids between the earth and Mars as a planet, as well as the three outermost planets Neptune, Pluto, and Uranus. But it does provide a dramatic illustration of Kepler's yearning for order—a yearning similar to that of the Pythagoreans who calculated planetary orbits on the basis of harmonious musical intervals. It also illustrates how difficult it must have been for anyone reading Kepler's work (like Newton!) to pick out the correct and relevant "laws" among all the incorrect and irrelevant ones.

Of course Kepler did not know the reasons or "causes" for any of his laws. His faith in a simple, ordered universe and his disbelief in the complicated epicycle system drove him to his discoveries. He employed the principle laid down earlier by Copernicus—that a theory's merits should not be based on its authoritative appeal, but rather on whether or not it can explain the physical phenomena in question with fewer arbitrary constructs and assumptions than a competing theory. By this standard, Kepler's ellipse certainly did qualify as a "better" theory of planetary motion than the old picture of perfect circles with numerous epicycles had.

During the middle part of the seventeenth century, after Galileo and Kepler, the stage was essentially set for the final master stroke of the Mechanical Revolution. One last step remained before Newton's synthesis of Galileo's terrestrial laws and Kepler's planetary laws could occur: the analysis of what accelerations were at work in the circular (or elliptical) orbits of the planets about the sun and of the moon about the earth. What made planets and moons travel in circles? Certainly not the "natural" circular motion of the quintessence—the celestial essence of the early Greeks clearly did not have a place in the Copernican model of the solar system.

Descartes had come closest to the answer. As we have seen, he realized that the natural motion of an object would make it continue in a *straight* line to infinity if no accelerations or retardations changed the motion. But what acceleration changes the motion of an object like the moon to make it move in a circle instead of in a straight line? Descartes said:

> Every body which is moved in a circular fashion tends constantly to move away from the circle that it describes.

He used the example of a rock tied to the end of a string and whirled by hand in a circular path:

> And we can even feel with the hand, while we make the rock turn . . . for it pulls and tightens the string to go directly away from our hand.

So Descartes knew that the string was pulling *on the rock*, trying to make it move or change its motion (accelerate) toward the hand (toward the center of the circle it described). For if this were not true—if, for example, the string broke—the rock would fly off in a straight line. The acceleration that forces the rock to change its motion from a straight line into a circular path is called **centripetal acceleration**; it is always directed toward the center of a circular path (see Figure 5-7).

Although Descartes realized the need for center-directed acceleration, it remained for Newton and Huyghens, independently, to obtain the magnitude of this acceleration. The magnitude of centripetal acceleration a_c, derived in Appendix 5 (pages 302–303), is

$$a_c = \frac{v^2}{r} \qquad \text{(5-1)}$$

where v is the **tangential** (orbital) velocity of the object moving in a circle and r is the radius of that circle. This formula is universally valid: Any kind of circular motion requires a centripetal acceleration of v^2/r (directed toward the center of the circle)

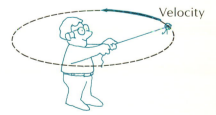

Velocity

5-7 When whirling a stone in a circle, the tension in the string keeps pulling in on the stone.

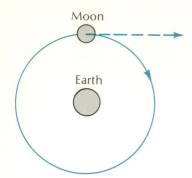

Moon

Earth

5-8 The moon would naturally move in a straight line at constant velocity to the right, as indicated by the dashed arrow, but, due to centripetal acceleration, it maintains a circular orbit around the earth.

to keep an object rotating with velocity *v* at a distance *r* from the center.

The principle of centripetal acceleration can be seen in Figure 5-8 which shows the moon in its 28-day orbit around the earth. At the instant shown, the moon is traveling to the right in the direction of the arrow and it would, of its own natural motion, like to continue moving to the right in a straight line. But it moves in a circle around the earth because the earth's gravity is continually "pulling" on it, like the string is continually pulling on the whirling stone. In other words, the moon is *falling* toward the earth at the same time it is traveling to the right. The amount the moon "falls" does not bring it any closer to the earth, but simply keeps it moving in a circular orbit at the same radial distance from the earth. With this result of centripetal force, the pieces of the puzzle given us by Copernicus, Galileo, and Kepler finally fit together into a new picture of the universe.

For Isaac Newton and the other scientists who believed in the Copernican universe, the burning questions were "Why does the moon orbit the earth in an ellipse?" and "Why do the motions of the planets obey Kepler's three laws?" Couldn't the motions of the planets around the sun and, better still, the motion of the moon around the earth possibly be related to Galileo's laws for the motion of terrestrial objects on the earth's surface? The answers to these questions were to deal a final blow to the Aristotelian universe and to mark the beginning of an entirely new mechanistic world-view—a physics and philosophy which, even today, has an enormous influence on our culture.

Questions

1 If a particle moves along a spiral path with constant speed, what do you know about the direction of its acceleration vector? Is it necessary to make any computations to answer this? If not, how do you know the answer?

2 Is it possible for a body moving in a circular path ever to have its acceleration vector point outward, away from the inside of the circle? Explain your answer.

3 Considering the rotation of the earth, can we say that objects fall truly vertically (toward the center of the earth)? Justify your answer.

4 Does the rotation of the earth affect the value of the gravitational acceleration of falling objects? If so, where on the earth would you expect the effect to be greatest? Least?

5 Kepler's first law states that planetary orbits are ellipses with the sun at one focus. Using a pencil, two pins or thumb tacks, and some string (see Figure 5-3), construct several ellipses. What happens to the ellipse if you move the pins closer together? If you add more string? If the two pins are placed together, what is the resulting figure? Is the circle a special kind of ellipse?

6 A comet moves much faster when it is near the sun than when it is far away from it. Use Kepler's second law to explain this fact.

7 Do you think it unusual that Kepler worked so long using the ellipse as an approximation to Mars' orbit before he realized the orbit was actually elliptical? Explain.

8 Contrast the methods Galileo and Kepler used in physics. Descartes and Kepler.

9 If the existence of Neptune had been known to Kepler, what would have happened to his planetary model based on the five regular solids?

Questions Requiring Outside Reading

1 What astronomical phenomena are in disagreement with Tycho Brahe's picture of the universe (Figure 5-2)?

2 What communication was there between Galileo and Kepler? What did each think of the other's work?

3 Would you consider Kepler a mystic? Explain.

4 What kind of interaction existed between Kepler and Brahe?

Problems

1 The radius of the earth's orbit is 93 million mi and the radius of Mars' orbit is 141.5 million mi. Using Kepler's third law, find the length of time in earth days that it would take Mars to orbit around the sun.

2 An automobile traveling at 60 mi/hr (88 ft/sec) goes around a turn with a radius of 400 ft. What acceleration does it experience in ft/sec²?

3 The radius of Jupiter's orbit is 7.78×10^{11} m and the orbital period of the planet is 4332.6 days. Find the centripetal acceleration of Jupiter in m/sec², assuming the orbit is a perfect circle.

4 Two satellites are in circular orbits around the earth; the ratio of their orbital radii is 3:4. What is the ratio of their orbital periods?

5 An earth satellite circles the earth once every 2 hr. Using Kepler's third law, determine the height of the satellite above the *surface* of the earth. (The *diameter* of the earth is 8000 mi. The radius and period of the moon's orbit are, respectively, 240,000 mi and 28 days.)

6 Greg whirls a ball in a circle on the end of a string 1 m long. If the speed of the ball is 3 m/sec, what is the value of the centripetal acceleration it experiences?

7 Throckmorton's girlfriend, Natasha, rides her bicycle at a speed of 10 m/sec. (a) What is the speed of a point on top of one wheel? (b) What is the speed of a point on the bottom of one wheel? (c) If one wheel has a radius of $\frac{1}{3}$ m, what is the acceleration of a point on the rim?

8 Throckmorton likes to drive his sports car around turns as fast as he can. If the maximum sideways acceleration his tires will permit

without skidding is 30 ft/sec², find the radius of the smallest turn he can manipulate (a) at 100 ft/sec (68 mph) and (b) at 150 ft/sec (102 mph).

9 Using the data from Problem 8, how fast can Throckmorton drive his car through a turn with a radius of 120 ft?

10 What happens to the value of the centripetal acceleration if the radius of a circle is (a) cut in half? (b) doubled?

11 What happens to the value of the centripetal acceleration if the speed is (a) doubled? (b) multiplied by 1.41?

12 What is the value of the centripetal acceleration required to hold the moon in its orbit around the earth? (Use the data for the moon given in Problem 5.)

13 What is the value of the centripetal acceleration required to hold the earth in its orbit around the sun? (The sun–earth distance is 93,000,000 miles.)

Newton and the Synthesis of the Mechanical Revolution

6

Isaac Newton was born in 1642, the year that Galileo died. Newton's grand synthesis was a universal system that not only encompassed all the physics of Galileo and Kepler, but also introduced a new picture of the universe—a mechanical picture—which has influenced western thinking down to the present day. The authority of Aristotelian physics and scientific methodology was finally to be broken and replaced by Newtonian Mechanics.

6-1

Introduction

Newton overcame the Aristotelian separation of physics and astronomy by showing that Galileo's set of "laws" for terrestrial motion and Kepler's "laws" for planetary (celestial) motion were both inevitable consequences of his own more general "laws." He revealed that *all* motions—constant velocity, constant acceleration, and even arbitrary, nonconstant acceleration—obeyed the same laws on earth as they did in the heavens. They could therefore be explicitly analyzed and calculated, making the motion of everything in the universe, in principle, completely predictable. Even the effects of friction or resistance could be accounted for empirically, although not all of their detailed "causes" or mechanisms were known at the time. Further, Newton invented a new kind of mathematics, the calculus, with which to perform these calculations.

Most of this work was done in a one and one-half year period when Newton was 23–24 years old. However, it was not until 20 years later that he developed several unfinished points and the

6-2

Newton's Contributions to Physics

*Isaac Newton
by Johan Vanderbank.*

The National Portrait Gallery, London

Principia was written. As Newton himself later recounted his early years,

> *I found the Method [of fluxions; i.e., the calculus] by degrees in the years 1665 and 1666. In the beginning of the year 1665 I found the method of approximating Series and the Rule for reducing any dignity of any Binomial into such a series [i.e., he had formulated the Binomial Theorem]. The same year in May, I found the method of tangents of Gregory and Slusius, and in November had the direct method of fluxions [the differential calculus], and the next year in January had the Theory of Colours, and in May following I had entrance into ye inverse method of fluxions [integral calculus]. And the same year I began to think of gravity extending to ye Orb of the Moon, and having found out how to estimate the force with which [a] globe revolving within a sphere presses the surface of the sphere, from Kepler's Rule of the periodical times of the Planets being in a sesquilaterate proportion of their distances from the centers of their Orbs I deducted that the forces which keep the planets in their Orbs must [be] reciprocally as the squares of their distances from the centers about which they revolve; and thereby compared the force requisite to keep the Moon in her Orb with the force of gravity at the surface of the earth, and found them [to] answer pretty nearly. All this was in the two plague years 1665 and 1666, for in those days I was in the prime of my age for invention, and minded Mathematics and Philosophy more than at any time since.*

NEWTON

Even as a youth, Isaac Newton was a mechanical wizard. Although considered a slow learner, at 12 he built an exact-scale working model of a windmill being constructed near his home. Unsatisfied with the windmill alone, he rigged up a treadmill as well and ran the mill with mousepower. Newton also studied the problem of flight at this early age, sailing paper kites at night with paper lanterns attached to them—a trick which caused a certain amount of consternation among witch-fearing peasants in the area. □ Newton was the kind of man who would study any field necessary in order to understand a problem. In 1663 he bought a book on astrology only to discover that he had to understand trigonometry in order to solve the astronomical problems it contained. Finding Euclid of little help, he delved into Descartes' *Geometry* and, although it was far beyond him, diligently worked through it line by line until he understood it. □ Newton was also interested in alchemy, a branch of science seeking to control the natural balance of the four elements (earth, air, fire, and water) in such a way as to convert lead into gold or to create a youth potion. He never succeeded at his efforts in this area. □ After graduating from Cambridge (without honors—he was weak in geometry), Newton spent two years in the solitude of the country to avoid the Great Plague which was then ravaging London and to work on scientific problems. During this time, he invented the reflecting telescope and began his work on the nature of light. □ His findings on the nature of light were the cause of several arguments with Robert Hooke, but these were miniscule compared to the near war about who invented the calculus that raged between Newton and his allies and the seventeenth-century German mathematician and philosopher Leibniz and his allies. Newton, whose work was published first, did note at the time that the two mathematicians had arrived at their conclusions independently, but he forgot his generosity when friends of Leibniz began hurling charges of plagiarism. Leibniz asked the Royal Society to defend his claim, but the Society's usual objectivity was somewhat affected by the fact that Isaac Newton was its president. The report, compiled largely by Newton's friends, was somewhat biased. □ Newton became involved in politics in 1687, when he led the fight against King James for freedom from political interference within Cambridge. Following this successful battle, the university elected Newton to its seat in Parliament, where, according to his friends, the only speech he ever made was to ask an usher to close the window. He did become acquainted with a number of important people in Parliament, including Montague—the Privy Councillor to the king—who, when placed in charge of reorganizing the country's monetary system, asked Newton to take the position of Director of the Mint—a highly-paid and well-respected job. (Voltaire claimed the appointment was actually due to the fact that Newton had a beautiful niece who eventually became Montague's mistress.) □ In his later years, Newton exhausted a great deal of effort using his scientific principles of mechanics to justify and clarify various historical events in orthodox religion. Shortly before his death in 1727, Newton said:

> I do not know what I may appear to the world; but to myself I seem to have been only a boy, playing on the seashore, and directing myself in now and then, finding a smoother pebble or a prettier shell than ordinary, whilst the great ocean of truth lay all undiscovered before me.

Newton was buried with great ceremony and honor in Westminster Abbey, where people came and still come to pay him homage. The well-known poet Alexander Pope summed up the feelings of many with the famous couplet,

> Nature and Nature's laws lay hid in night:/
> God said, Let Newton be! and all was light.

What revolutionary idea does Newton describe here? He "deducted" that the force which keeps the planets in their orbits around the sun and the moon in its orbit around the earth must be an *inverse square force*. Further, he used this hypothesis to compare the force necessary to keep the moon in its orbit with the gravitational force on the earth's surface and "found them [to] answer pretty nearly." With this statement, he demolished all of the Aristotelian methodology which had separated astronomy and physics for so long. To do this, first he introduced a new concept, **force**, and second he defined quantitatively and precisely the physical meaning of **mass**, the old concept of "content of matter" used only qualitatively by all physicists from the pre-Aristotelians through Kepler.

So Newton took the step that Galileo could not. He gave a "cause," a more fundamental explanation, for the acceleration that changes the velocity of an object by revealing (1) that such an acceleration is due to forces (pushes and pulls); (2) that the amount of such an acceleration is proportional to the amount of force applied to the object; and further (3) that the force required to accelerate an object is proportional to the mass (the "content of matter") of that object. Newton stated this "law" as

$$\vec{F} = m\vec{a} \tag{6-1}$$

where $\vec{F}$ is the force (a vector) and m is the mass. Now we need to define the units for F and m. From now on, we will use only the mks and cgs systems. In the mks system, the mass is measured in **kilograms** (abbreviated kg; 1 kg is about equal to 2.2 lb), and the force required to give a mass of 1 kg the acceleration of 1 m/sec^2 is called a **newton** (N). In the cgs system, the unit of mass is the **gram** (abbreviated g; 1 g equals 10^{-3} kg), and the force that produces an acceleration of 1 cm/sec^2 for a mass of 1 g is called a **dyne**. Since 1 m = 100 cm and 1 kg = 1000 g, we can see that 1 N = 100,000 dynes = 10^5 dynes.

The physical meaning of $\vec{F} = m\vec{a}$ is easy to understand. [Note that $\vec{a}$ is a vector quantity (see Chapter 4, page 46, and Appendix 4A, pages 297–99), and that force—like acceleration—has a direction and must be represented by a vector.] With a given constant force (say, 10 N), an object with a mass of 2 kg will accelerate at a constant acceleration

$$a = \frac{F}{m} = \frac{10 \text{ N}}{2 \text{ kg}} = 5 \text{ m/sec}^2$$

An object of twice the mass, 4 kg, will accelerate at half that acceleration or

$$a = \frac{10 \text{ N}}{4 \text{ kg}} = 2.5 \text{ m/sec}^2$$

In a similar way, by increasing the applied force, the acceleration increases proportionally. Thus the 4-kg object acted on by a force of 40 N will accelerate at

$$a = \frac{40 \text{ N}}{4 \text{ kg}} = 10 \text{ m/sec}^2$$

Example 6-1:

An automobile with a mass of 2000 kg moves with a constant driving force of 10^4 N. How far will the car travel in 10 sec? To find the distance traveled, we need to know the acceleration. Use $F = ma$ or $a = F/m$. If $F = 10^4$ N and $m = 2 \times 10^3$ kg,

$$a = \frac{10^4 \text{ N}}{2 \times 10^3 \text{ kg}} = 5 \text{ m/sec}^2$$

a *constant* acceleration, since the force producing it was given as constant. Then use

$s = \frac{1}{2} at^2 = \frac{1}{2} \times 5 \text{ m/sec}^2 \times (10 \text{ sec})^2$

or

$$s = \frac{5 \times 100}{2} m = \textbf{250 m}$$

Example 6-2:

What is the velocity of the automobile in Example 6-1 *after* the 10 sec acceleration? The velocity is obtained from $v = at$, with $a = 5$ m/sec^2 and $t = 10$ sec. So

$v = 5 \text{ m/sec}^2 \times 10 \text{ sec} = \textbf{50 m/sec}$

Example 6-3:

If the same automobile traveling at 50 m/sec is suddenly braked to a stop in 5 sec with a constant braking force, what is the amount of that force? First we have to find the acceleration or, in this case, the deceleration. Use $v = v_0 + at$, with $v = 0$, $v_0 = 50$ m/sec, and $t = 5$ sec. Then

$0 = 50 \text{ m/sec} + a \times 5 \text{ sec}$

$a = -\frac{50}{5} = -10 \text{ m/sec}^2$

Now, using this acceleration in $F = ma$, with $m = 2000$ kg,

$F = 2000 \text{ kg} \times (-10) \text{ m/sec}^2 = \textbf{-2} \times \textbf{10}^4 \textbf{ N}$

(The minus sign indicates that the force is directed opposite to the initial velocity.)

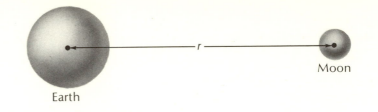

6-1 The gravitational attraction between the earth and the moon depends upon the distance *r* between their centers.

Earth

Moon

Returning to Newton's statement about the force required to keep the moon in its orbit, we find that, if the force acts along a line connecting the centers of the earth and moon (see Figure 6-1), then the **inverse square force** means that the force gets smaller as the separation of the centers increases.

The quantitative statement of this inverse square "law" is

$$F \propto \frac{1}{r^2}$$

where the symbol $\propto$ means "**is proportional to**." We cannot make an equality or an equation for F yet because F is also proportional to other things, as we will soon see. But we can reproduce the calculations that Newton used to find the force necessary to keep the moon in its orbit and to compare that force with the gravitational force at the earth's surface.

As we have seen, the moon describes a nearly circular orbit around the earth and so must have a centripetal acceleration pointing toward the center of its orbit and directed toward the earth. The moon would go flying off along a tangent to its orbit unless this centripetal acceleration were constantly forcing the direction of the moon's tangential velocity to change and execute a circular orbit. To put it another way, the moon is continually falling in toward the earth instead of moving along in a straight line with constant velocity.

According to Equation (5-1), the value of the moon's centripetal acceleration is $a_c = v^2/r$, where v is the tangential speed along its orbit and r is the distance of the moon from the center of the earth. The speed of the moon is the total distance it travels in one revolution divided by the total time the moon takes to complete one revolution. Thus $v = 2\pi r/t$, where $r = 240,000$ mi $= 240,000$ mi $\times 5280$ ft/mi $= 1.27 \times 10^9$ ft, and t (the time period of the moon's orbit) is about 28 days $= 28$ days $\times 24$ hr/day $\times 3600$ sec/hr $= 2.42 \times 10^6$ sec. Thus the tangential speed is

$$v = \frac{2\pi \times 1.27 \times 10^9 \text{ ft}}{2.42 \times 10^6 \text{ sec}} = 3.3 \times 10^3 \text{ ft/sec}$$

The centripetal acceleration is

$$\frac{v^2}{r} = \frac{(3.3 \times 10^3 \text{ ft/sec})^2}{1.27 \times 10^9 \text{ ft}} = 8.6 \times 10^{-3} \text{ ft/sec}^2$$

which is required to hold the moon in a circular orbit with a radius of 240,000 mi and a period of 28 days.

According to Newton's theory this centripetal acceleration is due to the gravitational attraction of the earth, which Newton hypothesized is inversely proportional to the square of the distance from the center of the earth. Thus if the acceleration due to the earth's gravity is 8.6×10^{-3} ft/sec^2 at a distance of 240,000 mi, it should be much greater at the surface of the earth—only 4000 mi from the earth's center. In fact, the inverse square hypothesis tells us it should be greater by a factor of $(240{,}000/4000)^2$ at the earth's surface over 8.6×10^{-3} ft/sec^2 at the moon's surface. This yields

$$a = \left[\frac{240{,}000}{4000}\right]^2 \times 8.6 \times 10^{-3} \text{ ft/sec}^2 = 31 \text{ ft/sec}^2$$

which, in Newton's words, "answers pretty nearly" to the gravitational acceleration measured experimentally as $g = 32$ ft/sec^2. (The fact that the two values are not exactly equal is due to our approximation of round numbers for both the radius of the moon's orbit and the moon's orbital period.)

In summation, Newton's assumption that the gravitational force produces an acceleration which falls off inversely as the square of the distance showed that both the motion of the moon and the downward, free-fall motion of objects on the earth could be explained by the *same* gravitational force. This single calculation broke the Aristotelian barrier dividing celestial and terrestrial phonomena and opened men's minds to a new way of looking at their universe.

6-3
Newton's Four "Laws"

Newton's first three "laws" give the "recipes" for calculating the motion (i.e., acceleration) of objects if the forces acting on them are known, or vice-versa:

1 Every object will continue in its state of rest (or of uniform motion in a straight line at constant velocity) unless an external force causes it to change that state.

2 An external force will change an object's state of motion by producing an acceleration according to

$$\vec{F} = m\vec{a} = m\frac{\Delta \vec{v}}{\Delta t} \qquad \text{(6-2)}$$

[This is derived from Equation (6-1) and Appendix 4A, pages 297–99.]

3 For every action (i.e., force) *on* an object, there is an equal and opposite reaction *by* the object upon the agent.

We note that the first law is merely a restatement of both Galileo's and Descartes' laws of inertia, with the word "force" replacing "acceleration." It is the second law that provides the connection between force and acceleration: They are proportional to one another in relation to the mass of the object. The mass *m* used in Equation (6-2) is called the **inertial mass**, since it determines the amount of resistance of an object to changes of motion. As we saw in the examples following Equation (6-1), for a given force *F*, a large mass will allow only a small acceleration or change in its motion. Such an object is said to have a large amount of inertia. On the other hand, a smaller mass will suffer a larger acceleration.

Newton's third law is particularly important if we wish to determine the motions of several interrelated objects. For example, if a man pushes *on a building* with a force *F*, then the building must be pushing back *on the man* with an equal force in the opposite direction. We will demonstrate the uses of these three laws later in Section 6-5.

Newton's fourth "law" is a quantitative description of one particular force occurring in nature.

4 The gravitational force that causes objects on the earth's surface to accelerate downward at 9.8 m/sec^2 and that causes the moon, due to gravitational attraction, to have the centripetal acceleration necessary to hold it in its orbit around the earth. A complete statement of this gravitational force is

$$F_{\text{grav}} = G\,\frac{m_1 m_2}{r^2} \qquad \text{(6-3)}$$

Where m_1 is the mass (in kg or g) of one body, m_2 is the mass of a second body, r is the distance (in m or cm) between the centers of the two objects, and F_{grav} is the resultant force (in N or dynes) with which the objects attract one another. *G* is a gravitational constant to make the units balance. If m_1 and m_2 are in kg and r is in m, then for F_{grav} to have the consistent units of N, $G = 6.67 \times 10^{-11}$ N-m^2/kg^2. For the cgs system, the value of the gravitational constant *G* is 6.67×10^{-8} dyne-cm^2/g^2.

The masses m_1 and m_2 in Equation (6-3) do not refer to the *inertia* of the bodies, as the mass in $F = ma$ did. Instead, these masses, which evidently produce the gravitational attraction, are

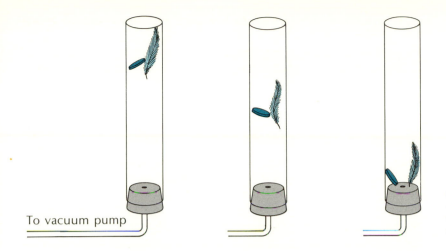

6-2 In the absence of friction or air resistance (as in a vacuum), a feather and a coin will fall at the same rate.

To vacuum pump

called the **gravitational masses**. The experimentally measured equality of the inertial mass and the gravitational mass is responsible for the equality in free-fall acceleration between light and heavy objects in a vacuum. For instance, in the absence of air resistance, a feather will fall at the same rate as a metal coin (see Figure 6-2).

We can show this by using $F = m_{inertial} \times a$ to find the downward acceleration a of a stone with an inertial mass of $m_{inertial}$. What is the force F that "causes" this acceleration? It is, of course, the gravitational force of attraction between the mass of the stone m_{grav} and the mass of the earth m_{earth}. This can be found from Equation (6-3):

$$F_{grav} = G \frac{m_{earth} m_{grav}}{r_{earth}^2} \qquad (6\text{-}4)$$

where r_{earth}, the separation of the stone from the center of the earth, is approximately equal to the radius of the earth. We can then use the gravitational force in Equation (6-4) for F in $F = m_{inertial} \times a$ to find a:

$$G \frac{m_{earth} m_{grav}}{r_{earth}^2} = m_{inertial} a \qquad (6\text{-}5)$$

Now assuming the equality $m_{grav} = m_{inertial}$ for the stone, the mass of the stone cancels and we find for a

$$a = G \frac{m_{earth}}{r_{earth}^2} = \text{constant} = g = 9.8 \text{ m/sec}^2 = 980 \text{ cm/sec}^2 \qquad (6\text{-}6)$$

the constant value g with which *all* objects will accelerate near the earth's surface where r_{earth} is approximately constant. Since the mass of the stone cancels in Equation (6-5), the free-fall motion of the stone is not dependent upon its mass. It is the same for *all objects*—feathers, coins, stones—in the absence of resistance or friction, as Galileo predicted.

It may seem odd that there are two distinct physical meanings of mass: (1) The inertial mass which characterizes the object's inertia or resistance to acceleration, and (2) the gravitational mass which characterizes the gravitational force on the object. It may seem even more odd that experiments show that these two seemingly different conceptualizations of mass are fortuitously equal to each other. Is this just a strange coincidence, or is there something more fundamental about mass that nature is trying to tell us? Two and one-half centuries after Newton's *Principia*, Albert Einstein was able to unify these two concepts in his General Theory of Relativity, as we will see in Chapter 16. For now, we will adopt the point of view of Newtonian physics, with the two different physical meanings for mass.

6-4

Derivation of Kepler's Laws from Newton's Laws

We have just seen how Galileo's laws of motion—constant velocity in a straight line for horizontal motion and constant acceleration downward at $a = g = 9.8$ m/sec$^2 = 980$ cm/sec^2 for vertical motion—naturally result from Newton's laws. Now we will see how two of Kepler's three "laws" of planetary motion can be derived from Newton's laws as well. The derivation of the first of Kepler's laws, stating that planets move in ellipses with the sun at one focus, can only be obtained from Newton's laws by using the calculus and is therefore beyond the mathematical level of this course. Those who are able to follow a simple derivation using the calculus are referred to any calculus-level textbook dealing with mechanics. We will merely state here that this derivation not only predicts ellipses for planetary orbits, but also obtains the orbits observed for the comets and for earth satellites.

Kepler's second and third laws can be easily derived with the use of only algebra and geometry. The third law states that R^3/T^2 is constant for all planets. We obtain this by using the centripetal acceleration required to hold the planet in its approximately circular orbit: $a_c = v^2/R$. Using Newton's second law, the centripetal force needed to keep the planet in orbit is

$$F_c = ma_c = m\frac{v^2}{R} \qquad \text{(6-7)}$$

Where does this force come from? What is its "cause?" For a ball whirled around on a string, the centripetal force is provided by the tension in the string pulling in radially on the ball. For a planet orbiting the sun, it is the gravitational attraction between the mass of the planet m and the mass of the sun M:

$$F_{\text{grav}} = G\,\frac{Mm}{R^2} \qquad\qquad (6\text{-}8)$$

where R is the radius of the orbit. So we use the gravitational force in Equation (6-8) to produce the required centripetal force in Equation (6-7), and set these two equations equal to one another. From this, with a little algebra (see Appendix 6, page 303), we obtain

$$\frac{GM}{(2\pi)^2} = \frac{R^3}{T^2} \qquad\qquad (6\text{-}9)$$

But all the factors on the left-hand side are constants independent of the planet. Thus Kepler's third law is a special case of Newton's laws. The value of the constant can be found knowing G and the sun's mass M. But most important, the "reason" for Kepler's third law has now become clear.

Kepler's second law, that a planet sweeps out equal areas in equal times, can be derived with the help of the diagram in Figure 6-3. Assume that a planet orbiting the sun takes the same amount of time to go from A to B as it does to go from B to C during what we will assume to be only a very small time interval. We are therefore trying to prove that the area SAB is equal to the area SBC. The planet's velocity is represented by the vector $\vec{v}_A$, tangent to the orbit at A. At B, the velocity is given by $\vec{v}_B$, tangent to the orbit at B. Now if this diagram is enlarged (see Figures 6-4 and 6-5), we see that $\vec{v}_A$ has been decomposed into two perpendicular components: $\vec{v}_{A\perp}$ which is perpendicular to the line SB and $\vec{v}_{A\parallel}$ which is parallel to this same line. Then

$$\vec{v}_A = \vec{v}_{A\perp} + \vec{v}_{A\parallel}$$

Likewise, $\vec{v}_{B\perp}$ is perpendicular to SB, $\vec{v}_{B\parallel}$ is parallel to SB, and $\vec{v}_B$ is tangent to the orbit at B, so

$$\vec{v}_B = \vec{v}_{B\perp} + \vec{v}_{B\parallel}$$

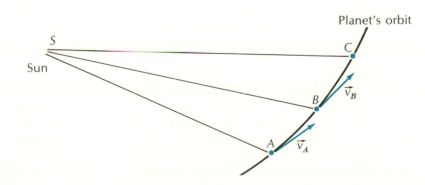

6-3 Motion of a planet in orbit around the sun. It takes as much time to go from A to B as it does to go from B to C.

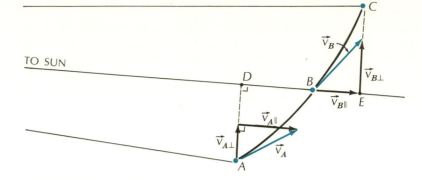

6-4 Proof of Kepler's second law.

Now Newton's second law, $\vec{F} = m(\Delta\vec{v}/\Delta t)$, states that the change in the velocity vector occurs only in the direction of the applied force. The gravitational force is an inverse square force, but its importance for our problem here is that its direction is **radial**, that is, along the line from the planet to the sun. Thus the gravitational force can only cause a change in that component of the velocity that is **parallel** to the sun-planet line *SB*.

The consequence of this radial attraction is that the component of the tangential velocity perpendicular to the sun-planet line *SB* ($\vec{v}_{A\perp}$ or $\vec{v}_{B\perp}$) is *not changed by the force of the gravitational attraction*. Thus, by Newton's first law, the perpendicular component remains unchanged from *A* to *C*. We call it a "constant of the motion." This means that the perpendicular distances *AD* and *CE* are equal if the time of transit from *A* to *B* is equal to that from *B* to *C*. If, as we have assumed, these times are very small, the arcs *AB* and *BC* can be approximated by straight lines, and then the triangles *SAB* and *SBC* have equal areas. This is true since both triangles have the same base *SB* and have equal altitudes, *AD* and *CE*. Since the area of any triangle is $\frac{1}{2}$ base $\times$ altitude, the areas *SAB* and *SBC* are equal and Kepler's second law—equal areas in equal times—has been demonstrated.

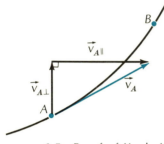

6-5 Proof of Kepler's second law.

6-5

Examples of Newtonian Mechanics

1 If you push on a building with a force of $\vec{F}$, why doesn't the building move with an acceleration of $\vec{a}$, following Newton's second law $\vec{F} = m\vec{a}$? This can be answered if *all* the forces *on* the building are considered. Since the building in fact does not move, the *net* force on it must be zero. Thus we conclude that your push of $\vec{F}$ *on the building* must be counteracted with an equal and opposite force of $-\vec{F}$ that is also acting *on the building*. The bolts and foundation that attach the building to the ground produce the required force ($-\vec{F}$) to cancel the force of your push.

(See Figure 6-6.) This building is said to be in **static equilibrium**: Equal, but opposite, forces are acting upon it; thus the *net* force on it is zero. By Newton's second law, the building's acceleration therefore must also be zero. Of course, if you push on the building hard enough, with the aid of a bulldozer, for example, the bolts or foundation holding it in place will break and will no longer be able to provide an equal force to cancel the applied force of the bulldozer. In such a situation, the building will start to accelerate and will usually collapse.

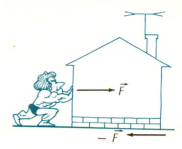

6-6 A man pushes a heavy object like a house and finds that it does not move. The foundation of the house pushes back on it with an equal and opposite force. The total force is thus zero.

2 If you try to apply your force to an automobile with its brake off, you will find that you must exert a rather large force before it begins to budge. Why doesn't it begin to move the first instant that you begin to push (Figure 6-7 a)? It has no bolts or foundation to provide an equal, but opposite, force as a building does. The answer here is that **friction** in all the moving parts of the car and between its tires and the road must be overcome before it will move. As in the case of the building, these frictional forces will cancel your applied force up to a certain amount (Figure 6-7 b). Beyond this amount, the constant maximum value of the frictional force can be overcome if you push hard enough (Figure 6-7 c). The force of friction $\vec{F}_{max}$ is normally found empirically. Repeated tests have shown that it depends upon the material composition of the frictional surfaces, the pressure of one surface on the other, the lubrication, etc.

Small force   No movement

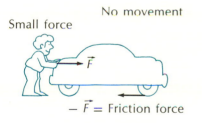

$-\vec{F}$ = Friction force

6-7a With a small applied force pushing the car the friction force will cancel the original force.

Medium force   No movement

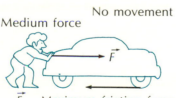

$-\vec{F}$ = Maximum friction force

6-7b A larger applied force is able to equal the maximum friction force.

Large force   Car accelerates

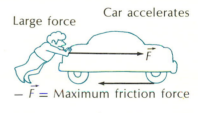

$-\vec{F}$ = Maximum friction force

6-7c An even larger force will overcome the maximum friction force and the car will accelerate.

3 If you place a block with a mass of m_1 on a horizontal, frictionless surface and push on it with a force of $\vec{F}$ (Figure 6-8 a), it will accelerate with $\vec{a} = \vec{F}/m_1$. If you place the m_1 block next to a larger block m_2 and apply the force $\vec{F}$ as before (Figure 6-8 b), what is the combined acceleration $\vec{a}$ of the two blocks? To analyze this problem, we have to abstract, considering each block separately and finding all of the forces acting on each one. Consider m_1 first.

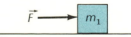

6-8a A single force $\vec{F}$ acts on a block of mass m_1.

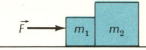

6-8b The single force $\vec{F}$ acts on both blocks m_1 and m_2.

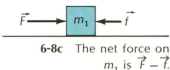

6-8c The net force on m_1 is $\vec{F} - \vec{f}$.

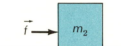

6-8d The force $\vec{f}$ acts back on m_2.

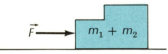

6-8e The force $\vec{F}$ acts on m_1 and m_2 as if they were glued together.

The applied force $\vec{F}$ pushes it to the right and an unknown force we will call $\vec{f}$ pushes it to the left at the same time because m_2 is pushing back on m_1. What is $\vec{f}$? You might think at first that Newton's third law—for every action there is an equal and opposite reaction—applies here so that $\vec{f}$ would be equal to $-\vec{F}$. But that cannot be right for, if $\vec{f} = -\vec{F}$, then the *net* force on m_1 would be zero and it could never accelerate.

So for now, we consider $\vec{f}$ to be unknown and write the equation for the acceleration of m_1 as

$$(\vec{F} - \vec{f}) = m_1\vec{a}$$

This equation is not of much help, however, since we do not know either $\vec{f}$ or $\vec{a}$. But we *can* find $\vec{f}$ by analyzing the block m_2 by itself because m_2 has only *one* force acting on it: $\vec{f}$ pushing it to the right. We know it must be $\vec{f}$ by applying Newton's third law: The force that m_1 exerts on m_2 is equal and opposite to the force that m_2 exerts *back* on m_1. So we write Newton's second law for the motion of m_2:

$$\vec{f} = m_2\vec{a}$$

Note that the acceleration $\vec{a}$ of m_2 is the same as the acceleration of m_1, since *they are in contact and must move together*. If we substitute the expression for $\vec{f}$ above in the earlier equation for the motion of m_1, we find,

$$\vec{F} - \vec{f} = \vec{F} - m_2\vec{a} = m_1\vec{a}$$

or

$$\vec{F} = m_1\vec{a} + m_2\vec{a} = (m_1 + m_2)\vec{a}$$

This result shows that the acceleration of the combination is simply the original applied force $\vec{F}$ divided by the *total* mass of the two blocks:

$$\vec{a} = \frac{\vec{F}}{m_1 + m_2}$$

Thus the motion is the same as if the two blocks had been connected all the time; we did not have to include the force $\vec{f}$ which acted between them. (If we had not obtained this last result, we would have ended up with a logical inconsistency in Newton's laws and Newtonian physics would have been in big trouble.)

4 Let us consider the meaning of weight next. **Weight**, which is a force, is *not* the same as mass (discussed in Section 6-3). A girl standing on the floor is in static equilibrium. The force of gravity, $\vec{F}_{\text{grav}} = m\vec{g}$, pulls downward on the girl due to the attraction of the earth's mass and an equal, but opposite, force $\vec{f}$ of the floor pushes *upward* on the girl's feet. These two forces cancel, yielding no net force and, therefore, no acceleration (see Figure 6-9 a).

The "weight" the girl feels is the force $\vec{f}$ with which the floor pushes up on her; this is the force measured by a set of scales. If the girl is motionless, her weight is just the same as the gravitational force pulling up on her. If she jumps out of an airplane, however, there is no force of a floor pushing up on her (although the gravitational force $F_{grav} = mg$ is still present and accelerates her downward with $g = 9.8$ m/sec^2). This condition, called "weightlessness," is experienced in free fall (see Figure 6-9 b).

We have an intermediate situation in an elevator moving downward from rest with an acceleration of $\vec{a}$. We feel less weight than normal and, indeed, a scale would register less than our usual static weight. In this case, the force of the floor $\vec{f}$ on the girl is *less* than the gravitational attraction $\vec{F}_{grav} = m\vec{g}$ (Figure 6-9 c). Consequently, there is a *net* force of $\vec{F}_{grav} - \vec{f}$ which does not cancel out to zero. The measured weight $\vec{f}$ is such that the net force $(\vec{F}_{grav} - \vec{f})$ on her produces the downward acceleration $\vec{a}$:

$$(\vec{F}_{grav} - \vec{f}) = m\vec{a}$$

or

$$(m\vec{g} - \vec{f}) = m\vec{a}$$

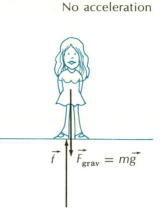

No acceleration

6-9a The gravitational force $\vec{F}_{grav} = m\vec{g}$ on a person standing on the ground is cancelled by the force of the floor pushing up on her.

Acceleration down

6-9b In free fall, neglecting air resistance, the only force is the gravitational force and a person accelerates at 32 ft/sec^2.

6-9c In an elevator accelerating downward, the gravitational force $\vec{F}_{grav} = m\vec{g}$ is greater than the force of the elevator floor $\vec{f}$ on the person.

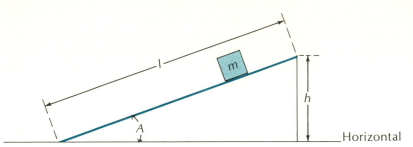

6-10 A block of mass *m* on an inclined plane.

If we transfer $\vec{f}$ to the right-hand side of the equation and $m\vec{a}$ to the left, we find the weight $\vec{f}$:

$$\vec{f} = m\vec{g} - m\vec{a} = m(\vec{g} - \vec{a})$$

Thus, if a person has a mass of 100 kg, his normal weight is

$$f = mg = 100 \text{ kg} \times 9.8 \text{ m/sec}^2 = \textbf{980 N}$$

If he now accelerates downward with $a = 2$ m/sec^2, his weight would be measured on a scale as:

$$f = m(g - a) = 100 \text{ kg} \times (9.8 - 2.0) \text{ m/sec}^2 = \textbf{780 N}$$

In a similar manner, when the elevator which was moving downward stops, the acceleration is upward and he has a greater than normal weight.

5 Now we will analyze the motion of a block on a **frictionless** inclined plane. (By frictionless, we mean that the plane cannot exert any force parallel to its own surface.) Let the length of the plane be *l* and let one end be a height *h* above the other, so that the plane makes an angle *A* with the horizontal. The block on the plane has a mass of *m*. What are the forces on it (see Figure 6-10)? First, it has a gravitational force $\vec{F}_{grav} = m\vec{g}$ directed downward. But the plane impedes this force by pushing back up on the block with a force we will call $\vec{F}'$; $\vec{F}'$ can only act perpendicularly to the plane assumed to be frictionless (Figure 6-11 a).

What is the magnitude of $\vec{F}'$? Since the block does not move in a direction perpendicular to the plane, it is evident that $\vec{F}'$ must be exactly cancelled by that part of the gravitational force $\vec{F}_{grav}$ which is perpendicular to the plane and which we will call $\vec{F}_\perp$. Thus, by vector addition, we can separate $\vec{F}_{grav}$ into two components: $\vec{F}_\perp$ acting perpendicular to the plane and $\vec{F}_\parallel$ acting parallel to or along the plane. In Figure 6-11 b, we replace $\vec{F}_{grav}$ with its two components, $\vec{F}_\perp$ and $\vec{F}_\parallel$, and look again at all the forces acting

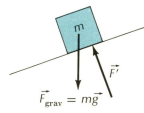

6-11a Two forces act on the block in Figure 6-10: the gravitational force $m\vec{g}$ and the force $\vec{F}'$ of the plane pushing up on the block.

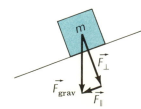

6-11b The gravitational force $\vec{F}_{grav}$ can be resolved into components: $\vec{F}_\parallel$ parallel to and $\vec{F}_\perp$ perpendicular to the plane.

on the block. Now in Figure 6-11 c we see that since $\vec{F}_\perp$ cancels $\vec{F}'$, there is no motion perpendicular to the plane. However $\vec{F}_\parallel$, the component of the gravitational force *along* the plane, is not cancelled by any other force. Thus, the block moves down the plane with an acceleration $\vec{a}$, obtained from Newton's second law (page 77) and due to $\vec{F}_\parallel$:

$$F_\parallel = m\vec{a}$$

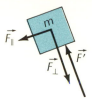

6-11c All of the forces acting on the block of mass m.

Now what is the magnitude of a? We find a by looking at the diagrams for the inclined plane (Figure 6-10) and for the force components $\vec{F}_\perp$, $\vec{F}_\parallel$, and $\vec{F}_{grav}$ (Figure 6-11 b); the right triangles in each diagram are similar triangles (see Figure 6-12) and, therefore, have proportional sides. The direction of $F_\perp$ is perpendicular to the inclined plane and F_{grav} is perpendicular to the horizontal; therefore, the angle A subtending h in the triangle on the left in Figure 6-12 is the same as the angle A subtending $F_\parallel$ in the triangle on the right, and the two triangles are similar. The sides of similar triangles are therefore proportional to each other; that is,

$$\frac{F_\parallel}{F_{grav}} = \frac{h}{l}$$

or

$$F_\parallel = \frac{h}{l} F_{grav}$$

The acceleration a of the block down the plane is then found by substituting $F_\parallel = ma$ and $F_{grav} = mg$ and cancelling m:

$$a = \frac{h}{l}g = \frac{h}{l}\ \textbf{9.8 m/sec}^2$$

Thus, we find the result originally proposed by Galileo: The acceleration down the plane a is in proportion with free-fall acceleration g as the height of the plane h is to its length l. We can therefore obtain very small values of acceleration by making h small, or we can obtain accelerations nearly equal to g by raising the plane so that h is nearly equal to l.

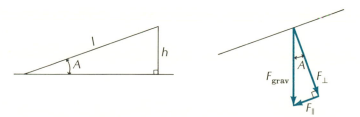

6-12 The angle A between the horizontal and the inclined plane (left) is the same as the angle A between F_{grav} and $F_\perp$ (right).

Example 6-4:

What is the acceleration of an object on an inclined plane 3 m long if one end of the plane is raised 10 cm above the other? The length of the plane is $l = 3$ m and its height is $h = 10$ cm $= 0.1$ m. Thus

$$a = \frac{h}{l}g = \frac{0.1 \text{ m}}{3} \times 9.8 \text{ m/sec}^2$$

$$= 3.27 \times 10^{-1} \text{ m/sec}^2 = \textbf{0.327 m/sec}^2$$

6-6

The Influence of Newtonian Mechanics

Many years elapsed after the publication of the *Principia* before a general acceptance of the Newtonian world-picture began to dominate the thinking of physicists, philosophers, mathematicians, and laymen alike. We can now see one of the reasons for this eventual acceptance. Copernicus' explanation of retrograde planetary motion eliminated many of Ptolemy's epicycles; Kepler's discovery of elliptical orbits eliminated the entire epicycle structure; Galileo's explanation of terrestrial motion as compound vertical and horizontal motions eliminated the many earlier, unsatisfactory theories of projectile motion; the development of ideas on inertia by Descartes and Galileo corrected many of Aristotle's incorrect arguments. However, Newton's four "laws" were able to explain *all* of these many previous "laws" or theories of motion using fewer constructs than his predecessors had to explain the same observations. Moreover, Newton's theories were so general, so sweeping in scope, that they were able to predict and explain current mechanical problems and astronomical observations. New stars and planets were "discovered" even before they were observed visually. When the motion of an observed star or planet was not in accordance with Newton's laws, astronomers postulated the existence of a nearby star or planet which could not yet be seen, but which was necessary to explain the irregular motion. When the postulated star or planet was finally observed, the occasion represented a great and spectacular triumph for Newtonian physics.

An industrial age was now possible. Newton's laws were indispensable in designing effective and efficient machines that would do the work of man—only the beginning of the social, economic, and political revolutions that became possible after Newtonian Mechanics began to rule the lives of men.

Questions

1   A covered wagon pulled by oxen is about to leave St. Louis. However, the oxen argue that it is pointless for them to even try to pull the wagon because, according to Newton's third law, it will exert

the same amount of force *on them* in the opposite direction. The two forces, say the oxen, would cancel one another and the net result would be two tired oxen standing in the same place. Should the pioneer forget about Oregon and set up a side-show to exhibit his talking oxen? Or should he reach for his whip? Explain where the forces must act to make the wagon move and the error in the oxen's reasoning.

2   If we work in a system where the fundamental quantities are *force, length,* and *time,* what are the dimensions of mass in that system? (*Hint:* Look at Newton's second law.) Do you think such a system would be meaningful? Why?

3   Explain simply whether or not the *numerical value* of the universal gravitational constant is dependent upon the size of the chosen standard of mass.

4   What is the ratio between the values of the gravitational acceleration at the surface of the earth and a distance of one earth-diameter in space from the center of the earth's gravity? How did you arrive at your answer?

5   What gravitational acceleration would you feel at the center of the earth? Explain.

6   Two teams, *A* and *B*, are having a tug-of-war. What does Newton's third law say about the force each team will exert on the other? What determines which team will win?

7   Can we determine the mass of a planet by observing its motion around the sun? Why?

8   Can we deduce the mass of a planet by studying its effect on the motion of another planet or planets?

9   Can we determine the mass of Jupiter by observing the orbital radius and period of one of its satellites? How? What relationship does your answer have to Kepler's third law?

10 According to Newton's third law, the earth exerts a force upon the sun which is equal and opposite to the force the sun exerts upon the earth. Then why doesn't the sun travel in an elliptical orbit about the earth?

Questions Requiring Outside Reading

1   Isaac Newton spent only a short part of his life working with physics. What else did he accomplish before and after he formulated the calculus, his three famous laws of mechanics, and his law of gravitation? Has your impression of Newton changed now?

2   What were Newton's ideas on the mechanism of gravitational attraction? Was this mechanism mentioned in his law of gravitation? Does the question of the mechanism of gravitational attraction have any meaning?

3   How is the universal gravitation constant *G* measured?

Problems

1 Throckmorton's car has a mass of 1000 kg. What force is required to give it an acceleration of 8 m/sec²?

2 How much does Throckmorton's car (in Problem 1) *weigh?*

3 A small rocket-propelled toy car has a mass of 20 g. Its motor exerts a force of 50 dynes. How fast does it accelerate?

4 A pool cue strikes a $\frac{1}{2}$-kg cue ball, exerting a force of 50 N for $\frac{1}{10}$ sec. (a) What acceleration does this produce in the cue ball? (b) How fast is the cue ball traveling when it leaves the cue?

5 A man whose mass is 10^4 g is accelerated upward in an elevator at a uniform rate of 100 cm/sec². (a) What is the man's normal weight outside of the elevator? (b) What does he weigh in the elevator?

6 Throckmorton has a string which will break when it is under a tension of 10^6 dynes. He ties a 1000-g lead ball to the string and whirls it in a circle with a radius of 100 cm. What is the tangential speed of the ball when the string breaks?

7 An object with a mass of 5 kg moves over a horizontal, frictionless surface. Starting from rest, it has a constant force of 60 N applied for 20 sec in a direction due north. After this time the force changes to a new force of 30 N directed due west for a time of 10 sec. What is the final velocity (magnitude and direction) of the object?

8 A block slides down a frictionless inclined plane 130 cm long. If one end of the plane is raised 50 cm above the other (a) what is the acceleration of the block? (b) If the block starts at the top of the plane, what is its speed when it reaches the bottom of the plane?

9 Natasha's car has broken down and Throckmorton is towing it home. If the mass of her car is 1500 kg and Throckmorton accelerates his car away from a stop sign at a rate of 2.5 m/sec², what is the tension in the towrope?

10 The radius of the moon is $\frac{1}{4}$ that of the earth and its mass is $\frac{1}{80}$ of the earth's mass. How much would a man who weighed 150 lb on earth weigh on the moon?

11 Looking through his telescope one night, Throckmorton sees a spherical asteroid with a radius of 2 km. If the asteroid is known to have a mass of 6×10^{10} kg, what is the acceleration due to gravitation at the surface of the asteroid?

12 If an electron (mass, $m = 9.1 \times 10^{-31}$ kg), traveling in a circular path of radius $r = 2.5 \times 10^{-10}$ m around a proton (mass, $M = 1.67 \times 10^{-27}$ kg), is held only by gravitational force, what must be the speed of the electron in its orbit? (*Hint:* The gravitational force must supply the centripetal acceleration of the electron.)

13 Use Newton's law of gravitation to find the gravitational force acting between one object of mass 10^3 kg and another of mass 5×10^3 kg when they are separated by a distance of 2.55×10^{12} m.

14 A binary star system contains two stars of equal mass, each traveling in a circular orbit around its common center of mass. If the stars are separated by a distance of 2.55×10^7 km and the gravitational

force on each has been calculated to be 4.0×10^{51} N, what is the mass of each star?

15 Throckmorton, star gazing again with Natasha, sees a giant, globular cluster of stars traveling in a circular orbit around the center of the galaxy. The mass of the cluster is 10^{38} g and the mass of the galaxy is 10^{44} g. If the radius of the cluster's orbit is 6.67×10^{20} cm, how long will it take the cluster to make one complete orbit around the galaxy?

16 The two blocks in Example 3 (pages 83–84 and Figure 6–8) are at the top of a frictionless inclined plane 5 m long, the top of which is 3 m higher than the bottom. What force accelerates the smaller block m_1 down the plane? What force accelerates the larger block m_2? What is the value of the acceleration for each block?

17 If the two blocks discussed in Problem 16 were in contact at the top of the plane, would they be in contact at the bottom? Why or why not? How fast would each block be moving at the bottom of the plane?

7 Energy and Momentum

7-1

Introduction

The Newtonian or mechanistic picture of the universe holds that all motion, from the operation of machines or the falling of an apple on the earth to the movement of the planets and stars, can be precisely explained by one or a combination of Newton's four laws. You may find it surprising that this view was not immediately accepted everywhere. Slowly, with the help of innumerable successes using Newtonian theory to predict astronomical and terrestrial phenomena, there eventually came to be as strong a *faith* in the Newtonian, mechanistic world-view (not just in science, but in most western thought, philosophy, and culture) as there had been previously in the Aristotelian, organismic universe.

The enormous scope and relevance of the Newtonian system were overwhelming, particularly in physics. There were very few basic discoveries or innovations for a century after the publication of the *Principia* due to the universality and the success of Newton's laws. A famous German physicist of the late nineteenth century, Ernst Mach, wrote

> *All that has been accomplished in mechanics since his [Newton's] day has been a deductive, formal, and mathematical development of mechanics on the basis of Newton's laws.*

Many extensions of Newtonian mechanics were to be made by mathematicians and physicists in the eighteenth and nineteenth centuries, but the results of all this work only demonstrated an increasingly broader applicability of Newtonian concepts.

The Newtonian principles for simple particles (Newton's four "laws") were thus extended to include the motions of more realistic solid bodies. The motion of a spinning top, a gyroscope, and other **rigid bodies** could then be precisely predicted by using Newton's laws for each point of the solid body and summing up all such points with the aid of the calculus Newton had developed for just this purpose. More formal, mathematical reformulations of Newtonian mechanics were also developed which made the solutions to complicated problems in mechanics possible with a minimum of effort. Many of these methods were retained and are still in use today. Perhaps the most important, the most relevant to our later study of physics, are the concepts of energy and momentum and their conservation "laws."

If we take Newton's second law $\vec{F} = m\vec{a}$ and replace the acceleration $\vec{a}$ with its definition as the change in velocity with time (see Appendix 4C, page 301),

$$\vec{a} = \frac{\Delta \vec{v}}{\Delta t}$$

we then find:

$$\vec{F} = m\frac{\Delta \vec{v}}{\Delta t}$$

Assuming that the mass is constant, we can place m within the "small change in" symbol:

$$\vec{F} = \frac{\Delta(m\vec{v})}{\Delta t} \tag{7-1}$$

The force is now defined in Equation (7-1) as the change in $(m\vec{v})$ with time. The product of mass × velocity is the **momentum**, and we give it the symbol $\vec{p}$. We see the usefulness of this new quantity if we assume that the force $\vec{F}$ is zero: then the change in momentum $\vec{p}$ must also be zero according to Equation (7-1). Thus, if there are *no external forces* acting on an object (or a group of objects), the momentum of the object (or objects) is constant and we say there is **conservation of momentum**.

For a single object that is moving in a straight line or that is in a state of rest, this principle is an obvious restatement of Descartes' laws for horizontal motion and of Newton's first law. However, if several objects are involved, the conservation of momentum principle is of much more significance. For example, if two balls are connected by a compressed spring, the total momentum of the "**system**" (as we will call the combination of

7-2

The Conservation of Momentum

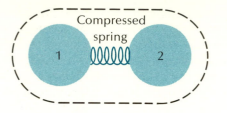

the two balls and the spring in Figure 7-1) is zero, since it has no velocity. When the spring is released and allowed to expand, the balls will begin to move in opposite directions (see Figure 7-2). However, this movement is not caused by an *external* force on the system; the force of separation is only the *internal* force provided by the spring. The momentum of the system is therefore conserved; since it was zero before, it also must be zero *after* the spring is released. Ignoring the small mass of the spring, this means that the momentum of ball 1 (the mass of ball 1 × its velocity) plus the momentum of ball 2 (again, its mass × its velocity) must equal zero. Let m_1 and m_2 represent the masses of ball 1 and 2, respectively, and v_1 and v_2 represent their corresponding velocities. Then, after the spring is released,

$$m_1 v_1 + m_2 v_2 = 0$$

or, for example, if we know v_1 and wish to know v_2:

$$v_2 = -\frac{m_1}{m_2} v_1$$

Thus, v_2 is in the opposite direction of v_1 (as indicated by the minus sign) and the magnitude of v_2 is to v_1 as the ratio of m_1 is to m_2.

Example 7-1:

Given $m_1 = 100$ g and $m_2 = 200$ g, if we observe m_2 to move to the right with a velocity of 10 cm/sec, what is the velocity of m_1? Use

$$v_1 = -\frac{m_2}{m_1} v_2 = -\frac{200 \text{ g}}{100 \text{ g}} \times 10 \text{ cm/sec}$$

$$= \mathbf{-20 \text{ cm/sec to the left}}$$

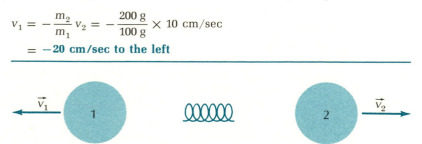

7-2 When the spring is released, the two balls fly apart in opposite directions.

7-3 A ball of putty with a momentum of $m\vec{v}$ hits a stationary cart and sticks.

Example 7-2:

A ball of putty with a mass of m is thrown against a cart of mass M and sticks to it. Will the cart move? If so, how fast?

Before hitting the cart, the putty has a momentum of $m\vec{v}$ (where m = mass and $\vec{v}$ = velocity; see Figure 7-3); the cart, at rest, has no momentum. The total mass of the (putty + cart) system after the putty sticks is $(M + m)$. By conservation of momentum, the momentum before $m\vec{v}$ equals the momentum after $(M + m)\vec{V}$, where $\vec{V}$ is the combined velocity of (putty + cart). Thus

$$(M + m)\vec{V} = m\vec{v}$$

or

$$\vec{V} = \frac{m}{(M + m)}\vec{v}$$

and the velocity of the cart is in the same direction as the direction the putty was thrown and **is smaller** than the putty's velocity **by the factor** $m/(M + m)$.

Example 7-3:

In place of the putty in Example 7-2, substitute a superball (which bounces with great elasticity) of the same mass. When thrown at the same cart with the same velocity as in Example 7-2, the ball bounces back from the cart. Will the cart move? If so, will its velocity be greater than, equal to, or less than its velocity $\vec{V}$ after being hit with the putty?

The total momentum of the (ball + cart) system before collision is $m\vec{v}$, just as it was in the (putty + cart) example. After collision, however, the ball is moving in the *opposite* direction with some velocity; we will call it $-\vec{v}_1$. Thus the momentum of *just the ball* after collision is $-m\vec{v}_1$. But the total momentum of (ball + cart) is conserved. So the total mo-

7-4 A superball with a momentum of $m\vec{v}$ hits a stationary cart and bounces back.

mentum after collision is that of the ball $-m\vec{v}_1$ plus that of the cart $M\vec{V}$, and the sum is equal to the momentum before collision $m\vec{v}$.

$$m\vec{v} = -m\vec{v}_1 + M\vec{V}$$

or

$$M\vec{V} = m\vec{v} + m\vec{v}_1 = m(\vec{v} + \vec{v}_1)$$

Solving for $\vec{V}$, we find

$$\vec{V} = \frac{m}{M}(\vec{v} + \vec{v}_1)$$

Compare this result with $\vec{V} = [m/(M + m)]\vec{v}$ obtained in the putty problem. The factor m/M for the ball is larger than $m/(M + m)$ is for the putty. Also the combined velocity $(\vec{v} + \vec{v}_1)$ for the ball is larger than $\vec{v}$ alone is for the putty. Thus **the cart's velocity** or momentum after being struck by the ball **is larger** than it was after being struck by the putty. The ball imparts an extra momentum $m\vec{v}$ to the cart just by *virtue of bouncing back.*

The conservation of momentum can be applied to even more complicated problems involving a combination of three, four, or many, many objects. A breakdown of this "law" has never been observed. Its usefulness and universality make it one of the several important laws of physics still used today.

7-3
The Conservation of Energy

Even before Newton, the **conservation of energy** principle was rooted in the energy concept, the *vis viva,* and in the arguments over the possibility or impossibility of constructing a perpetual motion machine. However, it was not until the middle of the nineteenth century that a clear, definitive enunciation of the principle was made independently by two *physicians,* Robert Meyer and Hermann Helmholtz. The latter was destined to change his field to physics and later to become one of the most famous physicists of his time. The energy conservation principle was first introduced to show that **heat energy**, which increases with temperature, is equivalent to the **mechanical energy** produced by the motion of the molecules of the heated object. Thus, as Meyer proposed, a vessel of water will increase in temperature when shaken. Our use of the conservation of energy will be restricted here to two kinds of mechanical energy: **kinetic energy** (of motion) and **potential energy** (of position).

Why are the laws of conservation of energy and momentum useful concepts if everything in the mechanistic world can be obtained from Newton's laws? The four laws are based on forces, and, to solve problems of motion, a detailed and exact knowledge of all the acting forces is essential. If some or all of the forces in a situation are unknown, then we cannot apply Newton's laws. But, as we saw in the preceding examples of conservation of momentum, we can still obtain information on the motion of objects, even when we do not know all of the forces acting upon them, by using the laws of conservation of momentum. It was not necessary to know exactly what force the compressed spring gave the two balls when it was released (Figure 7-2), since the force was not external. We could still solve for the motion of the balls. In the putty-cart and superball-cart examples, the forces of the putty and the ball hitting the cart were not known, but they were not needed to obtain information about the velocity of the cart.

The same situation is true of energy. We are able to avoid the necessity of a detailed description of forces by using conservation of energy, as we did in the momentum problems. It will be necessary to use both energy and momentum conservation to solve for the motion in most problems.

We *define* the change in energy as that quantity obtained by the product of a force $F \times$ the distance moved s, where F must be in the direction of (or parallel to) the distance moved. (If F is not in the direction of s, we use only that component of F which is parallel to s, as shown in Figure 7-5.) We find the **kinetic energy** KE of an object of mass m by beginning with Newton's second law for the force $F = ma$ and multiplying this by Galileo's equation $s = \frac{1}{2}at^2$ for the distance s. After a little algebra (see Appendix 7, page 303) in which the time t is expressed in terms of the velocity v, we see that the kinetic energy is half the product of the mass with the square of the velocity:

$$KE = \tfrac{1}{2}mv^2 \qquad\qquad (7\text{-}2)$$

The units of energy are seen from Equation (7-2) to be g-cm²/sec² in the cgs system (or, alternatively, dyne-cm, obtained by multiplying force $\times$ distance); this cgs energy unit is called an **erg**. Likewise, in the mks system, the energy unit kg-m²/sec² = N-m = **Joule**.

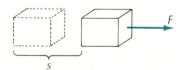

7-5 Energy is expended when an object is moved a distance s by a force F in the direction of the motion.

Example 7-4:

What is the kinetic energy of an automobile with a mass of 1000 kg traveling at a speed of 30 m/sec? $KE = \frac{1}{2}mv^2$, $m = 10^3$ kg, and $v = 30$ m/sec. Thus,

$$KE = \tfrac{1}{2}\,10^3 \text{ kg} \times (30 \text{ m/sec})^2 = \tfrac{1}{2}\,10^3 \times 900 \text{ kg-m}^2/\text{sec}^2$$
$$= \mathbf{4.5 \times 10^5 \text{ Joules}}$$

7-5

Potential Energy

The second "kind" of energy is energy of position or **potential energy** PE. Potential energy represents stored energy which, when released, will manifest itself in motion or KE. Examples of stored or potential energy are: the energy in a compressed spring; the energy obtained by moving a mass upward in the gravitational field g; the energy of a stretched rubber band; the energy in nuclear or molecular configurations that can be released in nuclear or chemical reactions. We will discuss only one example here: the gravitational PE near the earth's surface. Defining PE as force × distance, we find for the gravitational force $F_{\text{grav}} = mg$:

$$PE = F \times s = mg \times s = mgh \qquad \text{(7-3)}$$

(We use the conventional symbol h to represent height, rather than the symbol s for distance.) The units of PE are the same as those for KE (ergs and Joules).

Example 7-5:

How high must the automobile in Example 7-4 be raised for it to have PE equal to the KE of 4.5×10^5 Joules? The mass $m = 10^3$ kg and $g = 9.8$ m/sec^2. Using $PE = mgh$, we find:

$$h = \frac{PE}{mg} = \frac{4.5 \times 10^5 \text{ Joules}}{10^3 \text{ kg} \times 9.8 \text{ m/sec}^2} = 4.6 \times 10^1 = \mathbf{46 \text{ m}}$$

Finally, we derive the conservation principle in a manner analogous to that for momentum conservation. Since energy is the product of force × distance, we infer from Equation (7-2) that force is the change of energy with respect to distance, just as force is the change of momentum with respect to time. Thus, if the external forces are zero, then the energy of a system does not change. Energy is conserved; that is,

$$PE + KE = \text{constant}$$

which, for gravitational PE in Equation (7-3), reads

$$mgh + \tfrac{1}{2}mv^2 = \text{constant} \qquad \text{(7-4)}$$

Example 7-6:

A boy with a mass of 50 kg is on the top of a slide which is 2 m above the ground (see Figure 7-6). What are the boy's total potential and kinetic energies at the top of the slide? Figuring the height h from ground level, we find

$$PE = mgh = 50 \text{ kg} \times 9.8 \text{ m/sec}^2 \times 2 \text{ m}$$
$$= 980 \text{ kg-m}^2/\text{sec}^2 = 980 \text{ Joules}$$

$KE = \frac{1}{2}mv^2$ but, since the boy's velocity is zero at the top of the slide, $KE = 0$. Thus, his total energy of $KE + PE$ is just **980 Joules**

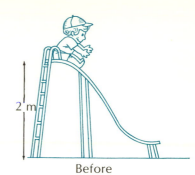

Before

Example 7-7:

If the boy in Example 7-6 slides down to the ground with no frictional (heat) energy loss, what are his total potential and kinetic energies now? After reaching the ground, his $PE = mgh$; but, at ground level, $h = 0$ and, therefore, his $PE = 0$. His KE is $\frac{1}{2}mv^2$. (We do not know v, but v is not needed since we will use the conservation of energy law.) The boy's total energy at the bottom of the slide equals his total energy at the top or

$KE = $ **980 Joules**

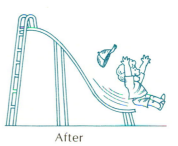

After

7-6 The boy has potential energy at the top of the slide. At the bottom, he has kinetic energy.

Example 7-8:

What is the boy's velocity at the bottom of the slide if his $KE = 980$ Joules? We set $KE = \frac{1}{2}mv^2 = 980$ Joules. Then, using the boy's mass of 50 kg,

$$980 \text{ Joules} = \frac{1}{2} \times 50 \text{ kg} \times v^2$$

or

$$v^2 = \frac{1960}{50} = 39.2 \text{ m}^2/\text{sec}^2$$
$$v = \sqrt{39.2} \text{ m/sec} = \textbf{6.25 m/sec}$$

The law of conservation of energy allows us to solve problems in mechanics in a simpler way. For instance, take Example 4-10 (page 49) where a ball is thrown straight up with an initial velocity of v_0. If we want to know how far up the ball travels, we must use both Equations (4-6) and (4-7) to eliminate the dependence on time and solve for the distance. But with the conservation of energy law, we can arbitrarily set $PE = 0$ at the position where the ball leaves the hand. Then the total energy is $KE = \frac{1}{2}mv_0^2$. The ball will stop at the top of its trajectory where $KE = 0$ (where the velocity is zero) and, due to conservation of energy, all the ball's energy will be gravitational potential energy PE = mgh at that point. So we can find the distance h the ball will rise by setting the total energy (PE only) at the top of the ball's trajectory equal to the total energy (KE only) at the bottom:

$$\tfrac{1}{2}mv_0^2 = mgh$$

We then solve for h and find that the distance the ball rises is

$$h = \frac{v_0^2}{2g}$$

Many other forms of potential energy exist in the world around us—an automobile tire filled with compressed air, a piece of wood held under water; the list of examples is endless. And as we will learn later, even the mass of an object in itself is energy according to Einstein's special relativity. The energy concept is one of the most useful and powerful tools in physics.

7-6
Angular Momentum

The concept of angular momentum arises from studying the rotation of rigid objects. Forces acting on an object can accelerate it in a linear motion according to $\vec{F} = m\vec{a}$, but forces can also set an object rotating. A baseball is thrown with a linear horizontal velocity, but a good pitcher adds spin to the ball to make it curve. The forces causing this spin or rotation produce a change in the ball's angular momentum with time. If the external rotational forces on a system are zero, the angular momentum does not change; it is conserved.

The **angular momentum** is given the symbol L and is defined as

$$L = mvr \qquad \text{(7-5)}$$

where m is the mass of the rotating object, v is the tangential velocity of the mass, and r is the perpendicular distance from the center (axis) of rotation to the velocity vector $\vec{v}$. The angular momentum of a baseball spinning about its own axis is difficult to calculate; since each mass point inside the baseball has a different tangential velocity and a different distance relative to the axis of spin, the angular momenta of all mass points would have to be added together to find the total angular momentum. This can be done only with the use of calculus, so instead we will illustrate angular momentum quantitatively with a simpler example: a small mass m on the end of a string whirling in a circle about a center C.

Example 7-9:

Suppose the mass = 100 g and the string has a length of 100 cm. If the mass is whirled until it has a velocity of 50 cm/sec, what will its angular momentum be then? $L = mvr$, $m = 100$ g, $v = 50$ cm/sec, and $r = 100$ cm. Thus,

$$L = 100 \text{ g} \times 50 \text{ cm/sec} \times 100 \text{ cm} = \mathbf{5.0 \times 10^5 \text{ g-cm}^2/\text{sec}}$$

Example 7-10:

If the string in Example 7-9 is shortened to 50 cm in such a way that no additional rotational force is given the mass, what would the velocity of the mass be then? Since angular momentum is conserved, $L = mvr$ is the same after the string is shortened as it was before. The mass remains constant, but r is reduced by $\frac{1}{2}$. The velocity must therefore increase by a factor of 2 to compensate. Thus, $v = \mathbf{100 \text{ cm/sec}}$.

We define the vector angular momentum $\vec{L}$ as having its length as its magnitude mvr and its direction *along the axis or rotation* perpendicular to the plane of rotation. $\vec{L}$ is normally defined in such a way that a counterclockwise rotation has an angular momentum vector pointed *up* as shown in Figure 7-7. The conservation of $\vec{L}$ applies not only to its magnitude, but also to its direction. A gyroscope or the spinning earth approximates this, since in each case the axis of rotation keeps pointing in the same direction in space. The conservation of angular momentum will not be pursued further here, but we will make use of it when we study the behavior of atoms and subatomic particles later.

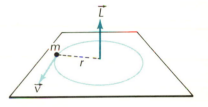

7-7 Angular momentum $\vec{L}$ is the product of $m\vec{v}$ and r. It occurs in the direction of the colored arrow, perpendicular to the plane where the rotation occurs.

Questions

1 A skyrocket bursts into two equal parts while still shooting vertically upward. Can we say anything about the directions of the velocity vectors of the two parts?

2 How does the law of conservation of momentum apply to a car accelerating along a highway? Momentum is apparently being created in one direction. Where is the momentum in the opposite direction which must cancel it?

3 Are Newton's third law and the conservation of momentum law consistent with one another? What is the relationship between the two?

4 Are definitions of the "fundamental" properties of physical objects and events—mass, length, time—more fundamental concepts than energy, momentum, and angular momentum?

5 If two objects collide and stick to each other, the collision will

conserve momentum and angular momentum, but kinetic energy apparently will not be conserved. What happens to the kinetic energy?

6 A man drives his car along the deck of a ferryboat toward a pier. Must the boat be tied up in order for him to drive his car onto the pier?

7 Do you find the existence of conservation laws surprising? Would you expect there to be properties of physical events which do *not* change?

8 Use the law of conservation of energy to explain why comets move much more rapidly when they are near the sun than when they are far away from it.

9 Now use the law of conservation of *angular momentum* to explain the movement of comets in relation to the sun. Is your answer compatible with your answer to Question 8?

10 Is it possible for two different objects to have the same kinetic energy, but different momenta? Give a numerical example.

11 Is potential energy itself defined uniquely, or is it only the difference in potential energy between points which is uniquely defined? What would happen to our physical calculations and predictions if a constant amount were added to the potential energy at every point?

Questions Requiring Outside Reading

1 Compare the early theories of "impetus" and of "impressed virtue" in Chapter 2 (page 21) with the laws of conservation of energy and momentum. Were these early theories *wrong* or were they merely incomplete? What did they lack?

2 According to Equation (7-1), force creates a change in linear momentum. What analogous quantity creates a change in angular momentum?

3 What were some of the important contributions made to heat energy theory?

4 Why do gases generally expand when they are heated?

5 What is meant by the "precession" of a gyroscope or top? Define "nutation."

Problems

1 A proton has a velocity of 2×10^6 cm/sec. If the proton's mass is 1.6×10^{-24} g, what is the magnitude of its momentum?

2 What is the speed of an electron if its mass is 9.1×10^{-31} kg and the magnitude of its momentum is 1.82×10^{-27} kg-m/sec?

3 Two steel balls in free space have a spring compressed between them and are held together by a cord. The masses of both balls are equal. The cord is severed by a meteorite, allowing the spring to push the balls apart. If the velocity of one of the balls relative to the spring (which remains stationary) is 5 m/sec, what is the speed of the other ball? (*Hint:* Momentum is conserved.)

4 Throckmorton is driving his car (total mass = 1000 kg) down the road at a speed of 40 m/sec. What is the car's kinetic energy?

5 What is the magnitude of the momentum of Throckmorton's car in Problem 4?

6 What is the kinetic energy of the proton in Problem 1?

7 When Natasha goes bowling with Throckmorton, she can roll her bowling ball (mass = 6 kg) down the alley at a speed of 5 m/sec. What is the ball's kinetic energy?

8 A moving ball of clay (mass = 5 g) strikes a stationary ball of clay (mass = 10 g) and sticks to it. After the collision, the velocity of the lump of clay (mass = 15 g) is 20 cm/sec. Find the initial velocity of the ball of clay with a mass of 5 g. (*Hint:* Momentum is conserved.)

9 A railroad car rolls along a siding until it strikes an identical railroad car that is standing stationary. After the collision, the resulting two-car train has a velocity of 0.5 m/sec. What was the initial speed of the striking railroad car?

10 An empty ore cart in the Leadville Mine rolls down a level track and strikes a full cart. After the collision, the two carts together move at a speed of 0.25 m/sec. If the initial velocity of the empty cart was 2 m/sec and its mass is 500 kg, find the total mass of the full cart.

11 If a toy car with a mass of 25 g has an energy of 2500 ergs, what is its velocity?

12 A small rocket motor pushes a toy rocket-car (mass = 1 kg) along a horizontal roadway, exerting a force of 3 N on the car. If the car starts from rest and travels a distance of 10 m *while* the rocket is firing, use the relationship between force and energy to find the kinetic energy of the car *and* its speed when the rocket has finished firing. Neglect any mass ejected by the rocket motor.

13 A ball with a mass of 2 kg is whirled on a string in a circle of radius 2 m with a velocity of 2 m/sec. What is the angular momentum of the ball?

14 If the radius of the circle in Problem 13 is suddenly increased to 4 m by releasing more string, what is the value of the new speed? (*Hint:* Angular momentum is conserved.)

15 A 500-kg race car rounds a turn of radius 20 m with a speed of 40 m/sec. What is its angular momentum?

8

The Electromagnetic Revolution

8-1

Introduction

In the eighteenth century, while Newton's laws were just gaining universal application and wide acceptance and while many of the earliest applications and extensions of Newtonian mechanics were being developed, some old and well-known phenomena were being reexamined in a quantitative way.

Even the early Greeks had observed that lodestone, a kind of rock, attracted metal objects made of iron, but did not attract nonmetallic objects or other nonferrous metals. The Greeks found this phenomenon, which we now call **magnetism**, very useful in navigation. When clouds obscured the stars and planets by which they normally charted their courses, the earth itself exhibited magnetic properties which caused a freely rotating magnetic needle to line up in an approximate north-south orientation. William Gilbert, physician to Elizabeth I, made the earliest quantitative studies of magnetism in the late 1500's. He tried unsuccessfully to use the results of his studies of terrestrial magnetism to "prove" the Copernican theory. Although detailed and exhaustive, Gilbert's work profited from neither the quantitative measurement nor the creative insight of his contemporaries—Galileo, Kepler, and Newton. As a result, quantitative measurements of magnetic effects were not to be successfully made until more than 200 years had passed.

The early Greeks also knew that the semiprecious stone amber had novel attractive properties which differed from those of lodestone. After being rubbed against a piece of fur, amber could attract pieces of paper or make a person's hair stand on end. This phenomenon, which we now call **static electricity**, was a mysterious force then. Both the properties of the magnetic force and static electricity were essentially ignored by the great minds of the Mechanical Revolution. Newton, while contributing greatly to optics and the study of light which were to constitute one branch of the Electromagnetic Revolution, for the most part ignored these magnetic and electric effects which today form another branch.

The attractive properties of amber were not studied seriously until the 1730's when an Englishman, Stephen Gray, discovered that a material's response to the electrostatic forces which provide amber's attractive power depended upon the composition of that material and not upon its color or any other qualities. He was able to distinguish good electrostatic **conductors** like the human body and metals from poor conductors like silk thread and glass. A few years later in France, Charles Du Fay accidentally discovered that there were *two* kinds of static electrical forces. One could be produced in the usual way: If amber were rubbed and then touched to some pieces of paper, the pieces would first be attracted to the amber, but then would be repelled once they touched the amber. However, when Du Fay rubbed a glass rod, he found that it *attracted* the same pieces of paper the amber *repelled* (see Figure 8-1). Du Fay proposed a two-fluid theory to account for these effects: One fluid causes the force effects of rubbed amber; a second fluid causes the force effects of glass. We can see the influence of Newtonian mechanics at work in Du Fay's model for electrostatic phenomena. Of course the "fluid" is weightless and cannot be detected directly by any of our senses, but it is a conveniently devised construct to explain the unusual electrostatic phenomena.

As soon as Gray's and Du Fay's studies were publicized by about the middle of the eighteenth century, widespread philosophic and scientific interest in electricity and magnetism began to grow. Benjamin Franklin in the United States was involved in many of these early experiments and, on the basis of his work, postulated that a one-fluid theory could explain Du Fay's findings. Franklin proposed that the attracting substance contained a

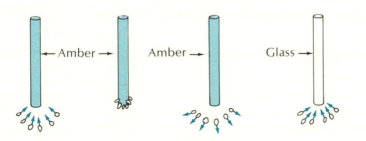

8-1 Pieces of paper are first attracted to and are then repelled by a charged amber rod. A charged glass rod attracts the paper just repelled by the amber.

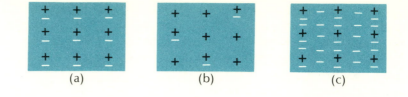

8-2

(a) A neutral object is equally charged, positively and negatively.

(b) A positively-charged object has a lack of negative charges.

(c) A negatively-charged object has an excess of negative charges.

(a)   (b)   (c)

"plenum of electrical fire" and that the repelling substance possessed a "vacuum of the same fire"; thus, only one fluid moved in electricity. Franklin's theory agreed with Copernican criterion: It was more economical than Du Fay's in its number of required constructs (one fluid versus two). Furthermore, it held essentially the same view of electrical phenomena as we do today: We believe that the **electrons**, the atomic units of negative electricity, are the only objects that move when a body is charged negatively; such a body has more of these electrons (Franklin's "plenum of electrical fire") than the heavier, fixed, positive **ions**. When a body is positively charged, it has a deficiency of these electrons (Franklin's "vacuum of the same fire"). We still apply Franklin's labels of "**positive**" and "**negative**" to these two conditions today (Figure 8-2).

8-3

Electricity: An Inverse Square Force

Several philosophers and physicists soon realized that the electrical force was similar to Newton's gravitational force in that it acted through an empty space (air or a vacuum) and attracted another object *without direct physical contact*. To determine what force law existed between these two objects, Auguste Coulomb performed a very delicate experiment (shown in Figure 8-3). He attached the lower end of a long vertical wire to a horizontal rod. At both ends of this rod were spheres of charge Q' and near each of these spheres was a similar sphere of charge Q. The force between the pairs of spheres caused the crossbar to rotate (in the case shown in Figure 8-3, to increase the distance between each pair of spheres). Coulomb found that the amount of the crossbar's rotation was proportional to the force between the pairs of spheres

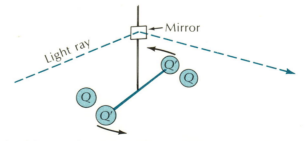

8-3 Coulomb's experiment to measure the electric force law.

and could be accurately measured by the amount of deflection given a light ray by a mirror attached to the wire. He published these results in 1785: The law of attraction and repulsion between two electrically charged objects is an inverse square force which diminishes as the square of the separation distance of the two charges increases, just like Newton's law for gravitation. Referring back to Equation (6-3), we know that the *gravitational force* F_{grav} is proportional to the product of the quantity of gravitational matter in one object (its *mass*) m_1 and the mass of a second object m_2 divided by the square of the distance between the two masses (see Figure 8-4), or:

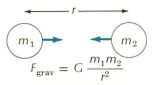

8-4 The inverse-square gravitation force between two masses, m_1 and m_2.

$$F_{grav} = G\,\frac{m_1 m_2}{r^2} \qquad (8\text{-}1)$$

Coulomb found that the **electrostatic force** is proportional to the product of the quantity of electrical "matter" or the **charge** of one object Q_1 and the charge of a second object Q_2 divided by the distance squared, or:

$$F_{electrostatic} = \frac{Q_1 Q_2}{r^2} \qquad (8\text{-}2)$$

This force law holds for point charges (that is, the charges Q_1 and Q_2 simply measure the number of excess electrons in negative charges or the number of missing electrons in positive charges).

The **repulsive force** on the point charge Q_2 due to the presence of Q_1 is represented in Figure 8-5 where Q_1 and Q_2 have the same charge sign (in this case, positive). Due to Du Fay's observations, an **attractive force** would result if Q_1 and Q_2 had opposite signs, one negative and one positive (also shown in Figure 8-4). The force on Q_2 due to Q_1 is always **radial** along the line connecting Q_1 with Q_2; if Q_2 were located directly above Q_1, the same statement would be true.

It is interesting to note that the inverse square nature of the

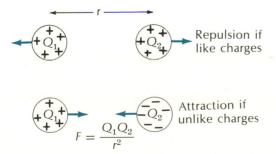

8-5 The inverse-square electrical force between two charges, Q_1 and Q_2.

electrostatic force had been previously hypothesized by Joseph Priestley, a famous English chemist who discovered the element oxygen and was a friend of Franklin. The following is taken from Whittaker's *"A History of the Theories of Aether and Electricity"*:

> [*Priestley was informed by Franklin that*] *he had found cork balls to be wholly unaffected by the electricity of a metal cup within which they were held; and Franklin desired Priestley to repeat and ascertain the fact. Accordingly, on 21 December 1766, Priestley instituted experiments, which showed that when a hollow metallic vessel is electrified, there is no charge on the inner surface (except near the opening), and no electric force in the air inside. From this, he at once drew the correct conclusion, which was published in 1767. "May we not infer," he says, "from this experiment that the attraction of electricity is subject to the same laws with that of gravitation, and is therefore according to the squares of distances; since it is easily demonstrated that were the earth in the form of a shell, a body in the inside of it would not be attracted to one side more than another?"*
>
> *This brilliant inference seems to have been insufficiently studied by the scientific men of the day; and, indeed, its author appears to have hesitated to claim for it the authority of a complete and rigorous proof. Accordingly, we find that the question of the law of force was not regarded as finally settled for eighteen years afterwards.*[1]

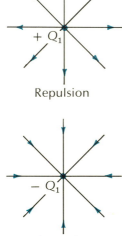

8-6 Two-dimensional lines of electric force due to charge Q_1.

In order to understand that Franklin's and Priestley's experimental results imply an inverse square law for the electric force, we will now introduce the new concepts of the **electrostatic field** and lines of electrostatic force. This **field theory** was to be developed and refined by Faraday and Maxwell[2] in the middle of the nineteenth century; it repudiates the Newtonian "empty space" picture by introducing a construct similar to Aristotle's "plenum": The electric field fills all space. This picture was to change again in the 1950's into a quantum field theory occupying an intermediate position between the Newton-Democritus "empty space void" and the Faraday-Aristotelian "plenum."

8-4
Lines of Force for Two Charges

Here we will devise a method to help remind ourselves of the direction of the force on any charge Q_2 due to its attraction to (or repulsion by) Q_1. We draw straight lines radially outward from Q_1, with outward pointing arrows to denote repulsion or inward pointing arrows to denote attraction. These **lines of force** are shown above in Figure 8-6. If Q_1 is positive, then, by convention,

---

[1] Edmund Whittaker, *A History of the Theories of Aether and Electricity* (Middlesex: Thomas Nelson & Sons, Ltd.), 1910. Used by permission.

[2] To be discussed in Chapters 10 and 12, respectively.

we use outward pointing arrows; if Q_1 is negative, we use inward pointing arrows. Figure 8-6 is a two-dimensional illustration. Actually, the lines of force extend in three dimensions (see Figure 8-7).

Now put the charge Q_2 at a distance r from Q_1 (see Figure 8-8). The arrows in the figure indicate the direction of the force on Q_2; the separation or spacing between the arrows tells us something about the magnitude of the force. We know the force is an inverse square; thus, the *increasing* space between the lines as r increases can be interpreted as a *decreasing* force. The number of lines emanating from Q_1 (Figure 8-9) is irrelevant; we could put in twice, three times, or a hundred times as many force lines if we wished. However, to be consistent, once we decide on the number of lines emanating from Q_1, then the number from a charge of twice Q_1, for example, should not be arbitrary, but twice the number of lines. This property of electric forces—that the forces add linearly—may seem intuitively correct, but, as we have learned from the Mechanical Revolution in the seventeenth century, intuition is not always right. The simple additive property of forces, **the principle of superposition**, is an experimentally verified fact both for gravitational forces and electric forces.

This principle can be expressed mathematically by defining the number of force lines N in terms of the charge Q_1. First we draw a sphere of radius r with Q_1 at the center (Figure 8-10). Now it is fortuitous that the surface area of a sphere is $4\pi r^2$ or proportional to r^2, while the force here is Q_1Q_2/r^2 or inversely proportional to r^2. Thus, the product of the force and the surface area A of the sphere of radius r will be *independent of the distance r*:

$$F \times A = \frac{Q_1Q_2}{r^2} \times 4\pi r^2 = 4\pi Q_1Q_2 \qquad \text{(8-3)}$$

This quantity does not depend on the separation r, but it is still dependent upon both Q_1 and Q_2; yet we said earlier that the lines of force emanated from only Q_1 and were independent of Q_2. If we divide the product of force × area by the magnitude of Q_2,

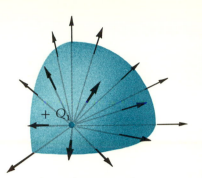

8-7 Three-dimensional illustration of lines of electric force due to charge $+Q_1$.

8-8 A charge Q_2 "feels" the electric force of Q_1.

8-9 The number of lines emanating from Q_1 is arbitrary.

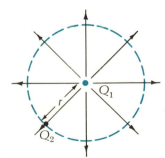

8-10 Imaginary sphere of radius r is centered on Q_1.

we obtain a quantity we will call N, the **number of lines** originating on a point charge Q_1:

$$N = 4\pi Q_1 \qquad\qquad\qquad\qquad \text{(8-4)}$$

This definition of N gives us an unambiguous mathematical description of the electric force. If the charge of Q_1 is doubled, then, according to the equation, N is also doubled. Thus, we see that Equation (8-4) implies a rule for force lines: Since N is defined so that it does not depend on the distance r, the total number of lines N from Q_1 will be the same at *any* distance, r or $2r$ or $5r$, and so on. So force lines are not lost or gained when traveling closer to or farther away from Q_1. *Force lines can begin or end only on charges.* This rule resulted from defining N as the product of an *inverse square force* (which decreases as the square of the distance increases) and the *area of the surrounding sphere* (which increases as the square of the distance increases), so that the result N is not dependent on distance any longer. Therefore, N must be the same for *all* distances.

8-5

Lines of Force for More Than Two Charges

So far, the rule of our method has been that if a charge Q_1 is increased or decreased, the number of force lines must increase or decrease proportionally. Now we ask, "What if we have *two* charges Q_1 and Q_3 which repel or attract our other charge Q_2?" (Instead of adding a charge to Q_1 as before, we will place the new charge Q_3 at a different point.) Assume that charges Q_1 and Q_3 are both positive (see Figure 8-11). If a charge were placed at point P, it would feel the sum of the forces $\vec{F_1}$ from Q_1 and $\vec{F_3}$ from Q_3. Because force lines $\vec{F_1}$ cross force lines $\vec{F_3}$, point P receives two forces $\vec{F_1}$ and $\vec{F_3}$ and a charge placed at P would not know which way to move. Since we already know that forces are

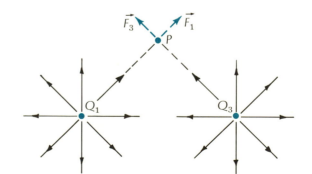

8-11 Two positive charges Q_1 and Q_3 produce two forces $\vec{F_1}$ and $\vec{F_3}$ on a charge at point P.

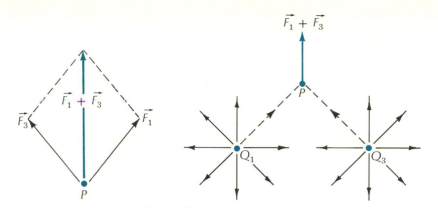

8-12 The two forces $\vec{F}_1$ and $\vec{F}_3$ at P are resolved into one force, the vector sum of $\vec{F}_1$ and $\vec{F}_3$.

vectors, we must use the *vector sum* of $\vec{F}_1$ and $\vec{F}_3$ for the *net* force on a charge placed at P (Figure 8-12). To find the net force at every point in space, we must first find the vector sum of the forces due to Q_1 and Q_3 everywhere. Since our first rule stated that force lines begin and end only on charges, the total number of lines leaving the combined charge of Q_1 and Q_3 is the sum of the individual number of lines from both Q_1 and Q_3. However, since it does not make sense for force lines to cross, the force lines from the two separated charges Q_1 and Q_3 will be curved as shown in Figure 8-13. An imaginary sphere (shown by the dashed line) will intercept N_1 force lines from Q_1 and N_3 force lines from Q_3, where $N_1 = 4\pi Q_1$ and $N_3 = 4\pi Q_3$ as before. This makes the total number of lines crossing this sphere merely the sum $N_1 + N_3$.

But what happens if Q_3 is placed a larger distance away from Q_1, beyond the radius r (Figure 8-14)? Force lines from Q_3 will

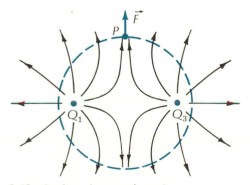

8-13 An imaginary sphere is constructed around both Q_1 and Q_3.

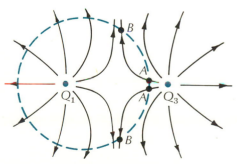

8-14 An imaginary sphere is constructed around Q_1 to the exclusion of Q_3.

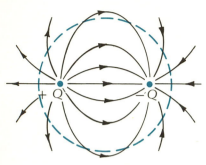

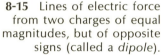

8-15 Lines of electric force from two charges of equal magnitudes, but of opposite signs (called a *dipole*).

enter the imaginary sphere around Q_1 (as at *A* points), but these lines will always leave the sphere (as at *B* points) because of our earlier rule that force lines begin and end only on charges. To calculate *N*, a line leaving the sphere at a *B* point cancels a line entering the sphere at an *A* point. Thus, no *net* lines of force from Q_3 leave or enter the imaginary sphere, giving us our second rule: The total number of lines leaving a sphere (or any other closed surface) is simply:

$$N_{\text{total}} = 4\pi Q_{\text{net}} \qquad (8\text{-}5)$$

where Q_{net} is the *net charge* enclosed by the sphere. This second rule is called **Gauss' law**

Example 8-1:

How many lines of force leave a sphere enclosing two equal, but opposite, charges $+Q$ and $-Q$? The answer is **zero**, since as many lines leave the sphere from $+Q$ as enter the sphere from $-Q$ (see Figure 8-15). (*Note:* The fact that the total number N_{total} is zero does *not* mean that the net force on a point *P* on the sphere due to the charges $+Q$ and $-Q$ is zero, since the number leaving does not cancel the number entering *at the same place on the sphere.*)

8-6

Lines of Force from a Uniformly Charged Sphere

Now we come to the first of our purposes for introducing lines of force: to show that the force on a point charge Q_2 due to a uniform, but extended, spherical charge Q_1 is the same as if the total charge Q_1 had been concentrated at the center of the sphere. Assume first that a total amount of charge Q_1 is **uniformly** distributed throughout a sphere. Due to the **spherical symmetry** of the problem, there can be no particular preferred direction around the sphere. The lines of force must look the same with the sphere in one particular position as they do if the sphere is rotated about any axis to a different position. The only possible direction of the lines of force is, therefore, **radial**—away from or toward the center of the sphere. A nonradial line would result in a preferred orientation of the sphere and would be contrary to the assumption of spherical symmetry. With the point charge Q_2 at a distance *r* from the center of Q_1 (Figure 8-16), how many lines of force *N* leave the imaginary spherical surface located at a distance *r*? Our second rule for lines of force, Gauss' law, tells us that

$$N = 4\pi Q_1$$

where Q_1 is the total net charge inside the large spherical surface of radius *r*.

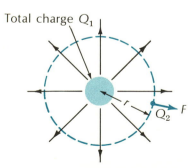

8-16 Lines of electric force from a uniformly charged sphere.

The area of this imaginary spherical surface is $4\pi r^2$ and, with the definition of lines of force $N = $ (force $\times$ area)$/Q_2$ from Equation (8-3), we can calculate the force on Q_2 due to the total charge Q_1:

$$F = \frac{Q_2 \times N}{A} = \frac{4\pi Q_1 Q_2}{4\pi r^2}$$

or

$$F = \frac{Q_1 Q_2}{r^2} \tag{8-6}$$

But this is the same force as we had for two point charges in Equation (8-2). Thus, the force on Q_2 here is the same as the force obtained if *all the charge Q_1 is concentrated at the center of the sphere,* since force equation (8-6) is *identical* to Equation (8-2) for point charges. This same result holds for gravitation: The force on a point mass m_2 due to an extended spherical mass m_1 is the same as if all the mass m_1 were concentrated at the center of the sphere. This is the result that Newton believed intuitively and proved in a different way by using the calculus. It is important to remember, however, that it is true only for an inverse square force where the total number of lines N is not dependent upon the distance r.

Next we will use the rules for lines of force to derive some simple properties of electrical conductors. First we need to define an **electrical conductor** as any material (almost any metal) which allows the free movement of electrical charges. We now believe these charges to be negative "conduction" electrons. The electrons in a **nonconductor** are not free to move because strong atomic and molecular forces hold them in one place, making the flow of electrical charges—or the "electric fluid," as Franklin would have called it—impossible. Whether a material is a conductor or a nonconductor depends both upon the structure of its individual molecules and the pattern that these molecules form in the material (as a crystal lattice).

First we will show that there can be no lines of force on a point charge Q_2 *inside* a conductor. This is easy. If there were lines of force, then a free point charge Q_2 would move. Since this is contrary to our definition of electrostatics—which deals only with charges at rest—no force lines can exist *inside* a conductor.

Secondly, there can be no free (excess) charges *inside* a conductor, a result which follows naturally from the preceding conclusion. If there were a free positive or negative charge Q, then force lines would begin or end on it. But since we have just shown that there cannot be any lines of force inside a conductor, no free charges can exist within a conductor either.

Thirdly, all the charges on a conductor must reside on its *outside,* since a charged conductor cannot have any charges *inside.*

Finally, we derive the result used by Priestley to infer an inverse square law for the electrostatic force given in Equation (8-2). Imagine a spherical conductor, a solid chunk of metal, with a total charge Q placed on it. As we already know, all of the charge must be on the outside of the metal and no charge or lines of force can exist inside it. Now imagine a small hole in the center of this chunk (Figure 8-17). The presence of the hole cannot affect any of our previous results. There cannot now be any free charges or lines of force either in the conductor or in the hole if there weren't any before the hole was made. If the hole is then enlarged until the conductor consists only of a thin spherical shell (or a shell of any shape) of a conducting metal, we still have not changed any of the earlier properties of the solid sphere. For the thin shell, the total charge Q still resides on its outside and there can be no lines of force inside the shell. Thus as Franklin and Priestley showed (Figure 8-18), when a small test charge was inserted into a narrow opening in a charged metal shell, there was no electrostatic force on the test charge because there were no lines of force inside the shell that had originated on or ended on the outside charges.

Priestley took this experimental result and, arguing backwards, inferred that the force between individual charges must be an inverse square force. Although his was not a direct proof (he still needed to show that other force laws of $1/r^3$, $1/r^4$, etc. could not lead to the same experimental result), the inference was brilliant.

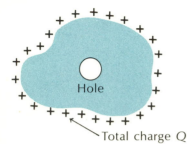

8-17 A conductor with total charge Q. All of the charge is on the outside of the metal and no charge or electric force exists inside the metal or inside any hole in the metal.

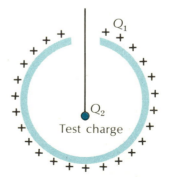

8-18 A hollow metal shell with total charge Q_1 which resides on the outside. A test charge Q_2 inserted inside the shell "feels" no force.

The basic equation of electrostatics

$$F = \frac{Q_1 Q_2}{r^2}$$

is called **Coulomb's law**. For this equation and for the rest of our study of electromagnetism, we will use the cgs system of units. In the force equation above, the units of force are in dynes and the distance r is in cm. We define the units of charge in terms of these basic units as: $Q = \sqrt{\text{dyne}}$-cm and, using $F = ma$, 1 dyne = g × cm/sec^2 and $Q = \sqrt{g \times cm}$ × cm/sec. We call this unit of charge a **stat-coulomb** (stat-C). [Note that by defining Q in terms of the basic units (cm, sec, and g), we eliminate the need for the constant in the electrostatic force equation that we had in the gravitational force equation $F_{\text{grav}} = Gm_1 m_2 / r^2$.]

Example 8-2:

A positive charge of 10 stat-C is located a distance of 20 cm from a negative charge of 4 stat-C. What is the force on the smaller charge? Use $F = Q_1 Q_2 / r^2$, where $Q_2 = 10$ stat-C, $Q_2 = -4$ stat-C, and $r = 20$ cm. Then

$$F = \frac{10 \times (-4)}{(20)^2} = -\frac{40}{400} = -\frac{1}{10} \text{ dyne}$$

(The minus sign indicates that F is an *attractive* force due to unlike charges.)

8-8

The Electric Field

In the preceding explanation of lines of force, we gave the lines from Q_1 an independent existence and used them when there were no forces (when there was no second charge Q_2). These lines represent what we call the **electric field** of Q_1: They give the direction of the force everywhere in space on a charge Q_2 if Q_2 is placed in the field. The magnitude of the force can then be obtained by multiplying the electric field $\vec{E}$ by the amount of charge Q_2 at that point:

$$\vec{F} = Q_2 \vec{E} \tag{8-7}$$

Thus we see that the electric field is simply the force per unit charge:

$$\vec{E} = \frac{\vec{F}}{Q_2} \tag{8-8}$$

and has the same vector direction as the force. From Coulomb's law, the electric field due to a point charge Q_1 or an extended sphere of charge Q_1 is

$$E = \frac{F}{Q_2} = \frac{Q_1 Q_2}{r^2 Q_2} = \frac{Q_1}{r^2} \qquad (8\text{-}9)$$

The units of E can be seen to be **dynes/stat-C** or **stat-C/cm²**.

Example 8-3:

A charge of 50 stat-C is placed in an electric field of 10^3 dynes/stat-C. The direction of the field is due West. What is the force on the charge? Use $\vec{F} = Q\vec{E}$, where $E = 10^3$ dynes/stat-C and $Q = 50$ stat-C. Then

$F = 50$ stat-C $\times\ 10^3$ dynes/stat-C

$= \mathbf{5 \times 10^4}$ **dynes directed due West**

To avoid confusion, we define the arrow of the electric field vector as the direction in which a *positive charge* would move. Thus a negative charge in an electric field will move along the field lines $\vec{E}$, but in the opposite direction (see Figure 8-19).

The mathematical definition of $\vec{E}$ given in Equation (8-9) is all well and good, but does the electric field really have an *independent* existence? Is there any physical effect due to a charge Q_1 which generates the field when a second charge Q_2 is not present? Since we only measure the field $\vec{E}$ by the force $\vec{F} = Q_2\vec{E}$ and the resulting acceleration on Q_2, the concept of the field itself seems unnecessary. However, it simplifies the analysis of many complicated problems, as we have already seen in connection with the force law for a spherically charged object. Furthermore, most physicists, following the tradition set by Faraday and Maxwell, consider electric and magnetic fields to be real physical quantities with independent existences, even when no charges are present

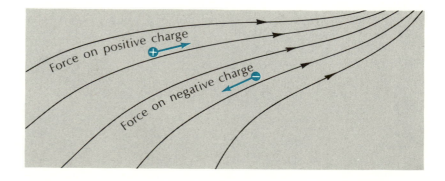

8-19 All space is filled with electric field lines. Positive charges move along these lines in the direction of the arrows; negative charges move in the opposite direction.

to measure the force effects of these fields. Thus, a single charge is assumed to fill up all of space with its electric field, and all space, instead of being the void Newton and Democritus suggested, is assumed to be filled with the electric fields of the countless numbers of charges in the universe.

But why isn't this same concept valid for the gravitational force? Newton did not hypothesize as to the origin of this force. He merely set down the force law $F_{grav} = Gm_1m_2/r^2$ and left the question of the mechanism of the force unanswered. In the *Principia* he states, "*Hypotheses non fingo,*" and even today the post-Newtonian world thinks of the gravitational force as an "action-at-a-distance" force. Newton himself was not happy with this definition; he believed that space had to have properties that somehow conveyed the force from one mass to another. However, he was either unwilling or unable to construct a hypothesis for a model of the gravitational force. Many, many physicists since Newton have tried to construct a satisfying gravitational field model with little success (except for Albert Einstein whose "Principle of Equivalence" will be discussed later in Chapter 16). The concept of the electric field represents the beginnings of a model for the "action-at-a-distance" forces between charges. As we will soon see, the magnetic field contributed even more to this model.

Questions

1 Can you think of any reasons why the quantitative study of motion and forces preceded the quantitative study of electricity?

2 What similarities *and* differences exist between electrical forces and gravitational forces?

3 Do you think that "action-at-a-distance" is or is not a reasonable way to describe electrical and gravitational interactions? What is an alternative? Explain.

4 How did the experiments of Franklin and Priestley lead to the conclusion that the electrical force varied as the inverse square of the distance separating the charges? (A diagram might be helpful in answering.)

5 Does Coulomb's law say anything about the *mechanism* of the electrical force? Does the electric field concept say anything about the mechanism?

6 Why do we need the constant G in the gravitational force law if no constant is needed in Coulomb's law?

7 What is the principle of superposition and what is its meaning in electrostatics? Can you think of an example where superposition does not hold? (You need not limit yourself to physics.)

8 Draw a diagram representing several charges surrounded by a spherical surface; now draw in lines of force through the surface. What happens to the number of lines of force passing through the

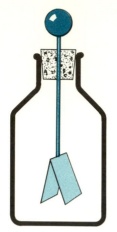

8-20 The electroscope.

surface if the size (radius) of the sphere is doubled? What happens if the number of charges is doubled?

9   Discuss the relationship, if any, between the concepts of the "electric field" and "lines of force."

10   Define a conductor and explain why metals are such good conductors. Why is there no electric field within a conductor?

11   You now know of two forces in nature, gravitational and electric. Can you find any reason to believe that additional forces do or do not exist?

12   In an electroscope (see Figure 8-20), two leaves of gold foil are connected to a metal rod which extends down from a metal ball. The apparatus is mounted in the mouth of a bottle. The gold leaves are protected inside the bottle; the ball remains outside. The gold leaves fly apart when the ball is touched by a charged object. Why does this happen?

13   Why can lines of force never cross one another?

14   A one-fluid theory suffices when we describe electrical phenomena in solids, but we must use a two-fluid theory to describe electrical phenomena in fluids (liquids or gases). Can you explain this? (*Hint:* Consider the motions of positive and negative charges in the respective cases.)

15   Why will a charge spread itself *uniformly* over the surface of a spherical conductor? Would you expect it to be spread uniformly over the surface of an *irregularly-shaped* conductor? Why?

16   After you comb your hair, your dry comb may pick up small bits of paper. What is happening? Would you expect this to be more noticeable on a wet or on a dry day?

Questions Requiring Outside Reading

1   Discuss some of the early theories of electricity. (Check Whittaker's *A History of the Theories of Aether and Electricity.*[1]) What features of these early theories, if any, relate to our present day understanding of electrical phenomena?

2   Henry Cavendish measured the inverse square law for the gravitational force by means of an instrument similar to Coulomb's (Figure 8-18). Cavendish was also able to determine the gravitational constant *G*, but had to make much more delicate measurements than Coulomb's to do this. Can you explain why?

3   What were some of Franklin's other contributions to electrical research?

4   Did Franklin's career as a scientist play an important part in his later influence on politics?

5   Describe some of the early "magic" or "side-show" tricks that made use of static electricity.

1 What is the magnitude of the force exerted on a 2 stat-C charge by a second charge of 3 stat-C if the two charges are 10 cm apart?

2 What is the magnitude of the force in Problem 1 if the charges are separated a distance of 100 cm?

3 A charge of 5 stat-C feels a force of 0.05 dynes due to a second charge located 100 cm away. What is the size of the second charge?

4 Two charges, separated by 20 cm, exert a force of 0.25 dynes on one another. What force will they exert on each other if they are separated by only 10 cm?

5 If two identical charges separated by 1 m (100 cm) exert a force on each other of 10^{-4} dynes, what is the size of each charge?

6 Two identical charges of 1 stat-C exert a force on each other of 10^{-2} dynes. What is the distance between them?

7 A charge of 3 stat-C feels an electric force of 4 dynes. What is the strength of the electric field at the location of the charge?

8 A charge experiences a force of 5 dynes in an electric field of 4 dynes/stat-C. What is the size of the charge?

9 If a charge of 0.12 stat-C is placed in an electric field of 5×10^{-3} dynes/stat-C, what force does the charge experience?

10 What is the strength of the electric field at a distance of 20 cm from a charge of 0.4 stat-C? (Give your answer in dynes/stat-C.)

11 What is the size of a charge if it produces an electric field with a strength of 3×10^{-4} dynes/stat-C at a distance of 8 cm from itself?

12 Three charges of 2 stat-C each are placed in a straight line with 10 cm of space between each of them (see Figure 8-21). The two outer charges are positive and the center charge is negative. (a) What are the magnitudes and directions of the forces on the two outer charges? (b) What force will the central charge experience?

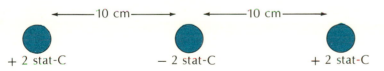

|←——10 cm——→| |←——10 cm——→|

+ 2 stat-C   − 2 stat-C   + 2 stat-C

8-21 Arrangement of charges for Problem 12.

9

Electric Currents

In the middle and late eighteenth century, the newly-discovered properties of electrostatics were used in a variety of ways. In addition to being employed as a curiosity in circus side shows and by medical quacks, scientific interest and experimentation in electrostatics continued to grow and a number of new technical advances were made. One of these was the "Leyden jar" (named after its place of discovery in Holland)—a plain glass jar lined on both its inside and outside surfaces with metal foil. Its purpose was simply to store electric "fluid" obtained by rubbing amber or glass. The modern version of the Leyden jar, called a **capacitor** or **condenser**, is widely used today in electronics. (As we know, the electric charge is not stored *in* the Leyden jar, but on the surfaces of the metal foils lining the jar.)

Another eighteenth-century device, the **electrostatic generator**, used mechanical energy (usually created by turning a crank on a machine) to continuously accumulate electric charge (obtained by rubbing or by a process called induction) and to store the charge in a Leyden jar. When the large positive and negative charges stored in the jar were allowed to neutralize each other, the electrons (Franklin's "electrical fire") moved from the negative pole of the machine or Leyden jar to the positive pole if a conducting metal wire was connecting the two poles. In addition, if sufficient charge were allowed to accumulate, the air (usually a poor conductor) became **ionized** between the poles and acted as a conductor itself for a split second. Charges passing between the poles to neutralize one another caused a loud "zap" and a lightninglike spark. (Today we call this movement of charges, either

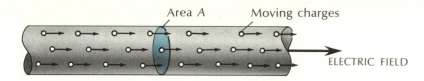

Area *A*   Moving charges

ELECTRIC FIELD

in the wire or in the air, an **electric current**.) But little more could be discovered about the properties of electric currents (moving charges) until more efficient means were invented to produce steady continuous currents, rather than the intermittent ones obtained from these electrostatic generators.

In 1780, just before Coulomb's experiment on the force law between charges, an Italian physician, Galvani, found that a dissected frog's leg twitched when it was touched at both ends by dissimilar metals. This phenomenon was originally called animal electricity and was thought at first to be only a property of animal tissue. In 1792, however, Volta showed that other moist materials sandwiched between two dissimilar metals could cause an electric effect and from that time the Voltaic pile, the predecessor of the modern battery, was used to generate electric currents from chemical energy.

The electric current is given the symbol *I* and is defined as the number of charges passing a given point per second or:

$$I = \frac{Q}{t} \tag{9-1}$$

The Unit of *I* is **stat-C/sec** or **stat-ampere** (stat-A). In Figure 9-1, charges (considered here to be positive) move in a metal wire. The number of charges that pass area A per second represent the current *I*. The charges move because of an electric field which exists along the wire; this field is controlled by the battery to which the wire is connected (see Figure 9-2). With the new phenomenon of electric current and with sources of continuous current flow, experiments were tried to determine the physical "laws" obeyed by currents. And, as we will see in the next few chapters, the study of currents led to several startling discoveries that were to cause considerable excitement in the scientific world. A new revolution was brewing—not only in the science of electricity, but also in the technology that science was to produce.

The first breakthrough happened quite by accident in 1820. Hans Christian Oersted, a professor of Natural Philosophy at the University of Copenhagen, had been investigating the actions static

9-2

Electric Current

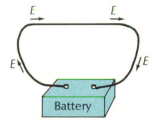

9-2 The electric field in the wire is provided by a battery.

9-3

Magnetism and Currents

Daguerreotype of Hans Christian Oersted, taken in the first ten years of photography.

The Burndy Library, Norwalk

electricity produced on a magnetic compass since 1807 with no results. Changes in compass orientation were known to accompany lightning bolts during electrical storms, but no laboratory duplication of this effect had ever been obtained. On this particular occasion, Oersted connected a wire to a Voltaic pile (battery) instead of to an electrostatic machine; he held the current-carrying wire over a magnetic compass at right angles to the compass needle (magnet) (see Figure 9-3). If there were to be a mutual force, Oersted expected it to cause the magnet to *align itself with the direction of the current in the wire* according to a possible force law similar to Coulomb's law or Newton's law of gravitation and felt that this effect could best be seen if the magnet was originally

9-3 Oersted's experiment. With the magnet (compass needle) perpendicular to the wire, there was no motional effect.

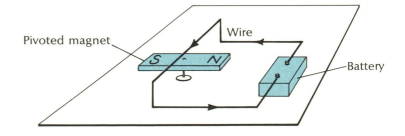

OERSTED Hans Christian Oersted was born in a small Danish town in 1777. His father ran the local drugstore and could not afford to send his sons away to school, so they were educated in their schoolless community by a wigmaker and his wife, the pastor, a surveyor, the mayor, and a local university student. This community-style, free-school education was effective enough to allow Oersted to enter the University of Copenhagen in 1793. ☐ He was impressed with the university, seeing it as a great home for thought and for the freedom to pursue truth. When he graduated, he became a lecturer and supplemented his income working for an apothecary. ☐ During his travels in Europe in 1800, Oersted met a physicist named Ritter who appears to have been a brilliant man. In a confused sort of way, Ritter combined physics and metaphysics, suggesting to Oersted that important electrical discoveries were always made in the years of the maximum inclination of the ecliptic. (True to this formula, Oersted eventually made his important discovery in such a year.) Ritter also gave some of his notes to Oersted which the latter remodeled into an essay for which he was awarded the annual prize of 3000 francs from the Institut de France. ☐ Oersted landed a professorship in physics in 1806 after having been rejected in 1803. He also managed a pharmacy and used its excellent experimental equipment to make a number of minor, but interesting, discoveries. ☐ In 1820, Oersted discovered the first relationship between magnetism and electricity while demonstrating the heating of platinum wire by electrical current to an advanced group of students. The needle on a compass placed near the wire swung away from North when the current was on. Whether or not this discovery was accidental is unclear; Oersted later claimed his actions were intentional, although at least one of his students was convinced that they were not. In any event, this first proof of an electromagnetic relationship was eventually to lead to the work of both Faraday and Ampère (discussed in Chapter 10) and to prepare the way for the development of the generator and the electric motor. ☐ After his discovery, Oersted devoted himself to education and lectured widely throughout the world. He founded a literary magazine and helped to establish a society to promote freedom of the press. He died in 1851 and was buried with special ceremony at the University of Copenhagen.

positioned perpendicular to the wire. However, he observed no motion—no effect at all. At the end of the lecture at which this negative experiment was performed, Oersted positioned the magnet *parallel to the wire*. Although a null result was expected since the magnet and wire were already aligned with one another, everyone was startled to see that when the current flowed the magnet rotated *away* from its orientation parallel to the wire and aligned itself *perpendicular* to the wire (Figure 9-4 b). If the current was interrupted, the magnet rotated back to its original position parallel to the wire (Figure 9-4 c). If the current was reversed and passed through the wire in the opposite direction, the magnet also rotated in the opposite direction, again to an alignment perpendicular with the wire. These results were unexpected, confusing, and astonishing. They established the first connection between magnetism and electricity and started a flurry

9-4a Oersted then positioned the magnet parallel to the wire; no current was in the wire yet.

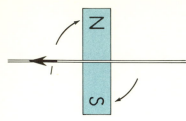

9-4b When current flowed through the wire, the magnet rotated to a position perpendicular to the wire.

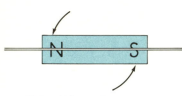

9-4c When the current was interrupted, the magnet rotated back to its original position parallel to the wire.

of new activity in the field of physics. Apparently the force law for the motion of a magnet due to an electric current was entirely different from the law for charges or masses. Two charges or two masses have a force directed along the line that joins them, but they do not seem to possess the circular or rotary kind of force that exists between a magnet and an electric current.

Of the many people who followed Oersted's experiment, André Ampère in France was one of the most actively interested. He knew that if a magnet was affected by a current, then, simply by Newton's third law of action and reaction, the wire carrying the current would be affected in turn by the magnet. From this he reasoned that a current in one wire might be effective in producing a force *directly on* another current-carrying wire without the intermediary of the magnet. Ampère experimented with this effect and, one week after learning of Oersted's result, he demonstrated that a force indeed existed between two electric currents. Surprisingly, in view of the strange circularlike force between a magnet and a current, this **current-current force** was a straightforward inverse square force. In Figure 9-5, two wires in two complete battery **circuits** are carrying currents I_1 and I_2, respectively. Ampère derived a force law from his findings which states that the force between a *current element* of circuit (1) e_1 and a current element of circuit (2) e_2 is simply the product of the two current elements e_1e_2 divided by the square of the distance r between them.

Ampère's law is written:

$$F_{\text{currents}} = \frac{e_1e_2}{r^2} \tag{9-2}$$

The **current element** e_1 is defined as the product of the current I_1 and a very short length of the circuit wire ℓ_1 (small enough so that ℓ_1 is approximately a straight-line segment and the distance r is approximately constant). This law (diagrammed in Figure 9-6) holds only if ℓ_1 and ℓ_2 are parallel to one another. If they make an angle with each other, then we must use the amount (compo-

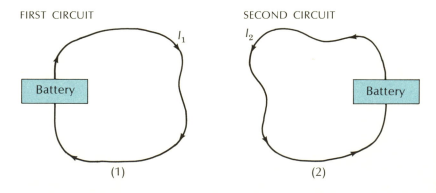

9-5 Two complete electric circuits.

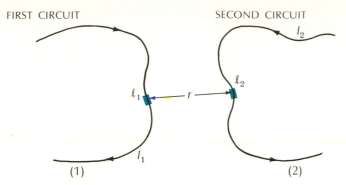

FIRST CIRCUIT   SECOND CIRCUIT

(1)   (2)

9-6 There is an inverse-square force between each current element of circuit (1) due to each element of circuit (2).

nent) of ℓ_1 in the direction of ℓ_2. For instance, if ℓ_1 and ℓ_2 are at right angles, there is no component of ℓ_1 in the direction of ℓ_2 at all and the force contribution is zero. Ampère also found the force to be *attractive* if I_1 and I_2 were in the *same* direction and *repulsive* if they were in *opposite* directions.

The total force between the circuits is obtained by adding up all the forces on ℓ_1 due to each length ℓ_2 of circuit (1) (Figure 9-6) and then adding up all the forces from the elemental lengths ℓ_2 for the complete circuit (2). This can only be performed in most cases by using calculus. The total force obtained this way can be a rather complicated expression, depending upon the geometrical arrangement of the wires in the two circuits. For the special case of two very long parallel wires (see Figure 9-7) the forces from the various current elements can be added together to give a simple result

$$F_{\text{currents}} = \frac{2I_1I_2\ell}{r} \qquad\qquad (9\text{-}3)$$

where ℓ is any given length of either wire (1) or wire (2). If ℓ refers to wire (2), then the force resulting from Equation (9-3) is the force felt by each length of wire (2) due to the long wire (1) a distance r away.

For the units of the currents I_1 and I_2 in Ampère's force equation, we again use the cgs system with force in dynes, distance

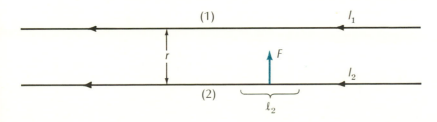

9-7 Equation (9-3) gives the force between two long wires carrying currents.

in cm, and time in sec. Since ℓ_1 and ℓ_2 both have units of cm and r^2 has units of cm², the force in dynes must have the same unit as the product of the two currents $I_1 \times I_2$. Consequently, the natural unit of current I for Ampère's law is $\sqrt{\text{dyne}}$. Now the current is charge/time or $I = q/t$; if I has natural units of $\sqrt{\text{dyne}}$, then the charge q used in interactions between *currents* has the unit $\sqrt{\text{dyne}} \times$ sec. Note that the symbol for charge in current—current forces (Ampère's law)—is a lower case q, whereas the charge for electrostatic forces (Coulomb's law) is represented by a capital Q. Are these two charges the same? Referring back to Coulomb's law (page 115), the units of Q are $\sqrt{\text{dyne}}$-cm. Thus the electrostatic charge Q cannot even be the same physical quantity as the electromagnetic charge q because it has different units! Comparing Q with q, we see that the two differ by a ratio which has the units of cm/sec—a *velocity*. We will explore this fact later and, for now, use the unit stat-coulomb (stat-C) for Q in Coulomb's law (electrostatics) and **ab-coulomb** (ab-C) for q in Ampère's law (electrodynamics). The corresponding current I is called an **ab-ampere** (ab-A).

Example 9-1:

A current of 10 ab-A flows in a long wire placed 15 cm from a parallel long wire with a current of 60 ab-A in the opposite direction. What is the force on a 5-cm length of the second wire? Using Equation (9-3), $F = 2(I_1 I_2 \ell_2 / r)$ with $I_1 = 10$ ab-A, $I_2 = 60$ ab-A, $\ell_2 = 5$ cm, and $r = 15$ cm. Then

$$F = \frac{2 \times 10 \text{ ab-A} \times 60 \text{ ab-A} \times 5 \text{ cm}}{15 \text{ cm}}$$

$$= 400(\text{ab-A})^2 = \textbf{400 dynes in a direction } \textit{away} \textbf{ from the first wire}$$

9-4

Electric Currents as Moving Charges

Currents result when an electric field is set up in a metallic wire with a battery. The electric field or force set up by the battery is directed along the wire, causing the free conduction electrons to move and an electric current to begin. Normally this electric force would accelerate the electrons in accordance with Newton's second law. Due to many collisions of the electrons with the stationary atoms of the wire, however, the **conduction electrons** actually move with an approximately constant velocity, called the **drift velocity**, when the battery produces the electric field. Thus the force described by Ampère's law is actually a force between the moving electrons in one wire and those in a second wire.

What if we have a charge like an electron moving in free space and not confined to a metal wire? Will there be a force between the free charge and a nearby current of conduction electrons within a wire? Yes, the physical picture is the same as it would be for the force between two currents in two wires. The only difference is that one current, the moving electron, is not constrained—it does not have to move along a wire.

In Figure 9-8, we show a charge q moving to the right with velocity v. The charge is near a wire carrying a current I_1 parallel to v. The force on the charge q is still given by Ampère's law, Equation (9-2):

$$F_{\text{currents}} = \frac{e_1 e_2}{r^2}$$

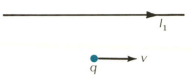

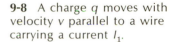

9-8 A charge q moves with velocity v parallel to a wire carrying a current I_1.

As before with Ampère's law, the current element e_1 within the wire is the product of the current I_1 with a small enough length of the wire ℓ_1 ($e_1 = I_1\ell_1$) so that the distance r from the charge to the current element is approximately constant. The current element for the moving charge is $e_2 = I_2\ell_2$. Recalling that the definition of current I was the charge flowing past a point per unit time or $I = q/t$, we can use this expression for the current in the current element:

$$e_2 = \frac{q}{t}\,\ell_2 = q\,\frac{\ell_2}{t}$$

But the small interval ℓ_2 divided by the time t is the particle's velocity v. Thus we have found that the current element for a moving free particle is

$$e_2 = qv$$

the product of the charge (in electromagnetic units, ab-C) and the velocity. The expression for the force on a moving charge due to current element e_1 is then

$$F_{\text{currents}} = \frac{I_1\ell_1 qv}{r^2} \tag{9-4}$$

If we vectorially sum up the forces from each of the current elements along a long straight wire, we find that the total force on the charge q is still proportional to qv and is still directed toward the wire (if the charge q is of the same sign and is moving in the same direction as the electrons in the wire are). We will always state conventionally that the current flows in the same direction as the electrons move. The sum of all the current elements $I_1\ell_1$ over the entire wire divided by r^2 can be performed via the calculus. Drawing an analogy to Equation (9-3), we find

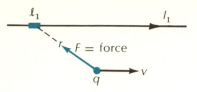

9-9 The force on q arises from contributions from each current element of the wire.

that the force on the charge q due to the current I in a long straight wire is

$$F_{\text{currents}} = \frac{2Iqv}{r} \qquad (9\text{-}5)$$

when q has a velocity parallel to the wire and is separated from the wire by a distance r. (See Figure 9-9.)

Example 9-2:

What is the force on a positive charge of 20 ab-C which is moving at a velocity of 5×10^8 cm/sec parallel to a long wire carrying a negative current in the same direction at the velocity of 10^3 ab-A? The distance from the charge to the wire is 2 cm. Use $F = 2Iqv/r$ with $I = 10^3$ ab-A, $q = 20$ ab-C, $v = 5 \times 10^8$ cm/sec, and $r = 2$ cm. Then

$$F = \frac{2 \times 10^3 \text{ ab-A} \times 20 \text{ ab-C} \times 5 \times 10^8 \text{ cm/sec}}{2 \text{ cm}}$$

$$= 10^{13} \textbf{ dynes directed away from the wire}$$

9-5

More about Forces on Moving Charges[1]

A force vector is directed along a line r between a current element $e_1 = I_1 \ell_1$ and a current element $e_2 = I_2 \ell_2$ (or $e_2 = qv$) if both elements are part of a continuous or complete circuit. However, a moving charge with current element $e_2 = qv$ does not constitute a complete circuit. Experiments tell us that the force law of Equation (9-5) is still valid in this case, but the direction of the force on the charge q due to each current element e_1 has a first contribution along r and a second contribution *along the direction of the current element e_1* (that is, along ℓ_1). The second contribution happens to cancel out when all the current elements of a complete circuit are added up.

This second contribution also cancels in the case considered above when $e_2 = qv$ is due to a moving charge traveling parallel to the current in a long, straight wire. It does not cancel in other cases where charges are moving toward or away from the wire. The magnitude of this additional component of force is still given by Equation (9-4) if v and r are parallel to one another and the direction of the force is along the wire or current element. If v and r make an angle with each other, we must use the component of v in the direction of r. If v is perpendicular to r, then this force contribution is zero.

[1] Section 9-5 represents material of greater difficulty which *can* be omitted without any great loss to the student in terms of overall understanding. The section has been included here for the sake of a completely accurate presentation.

For example, an electron moving at right angles away from a long, straight, current-carrying wire has no force contributions of the first type along the line r between the current element e_1 and the charge (see Figure 9-10). This is because the current element $e_2 = qv$ is perpendicular to all the current elements $e_1 = I_1 \ell_1$ in the long wire. However, the second force contribution in the direction of I_1 or ℓ_1 is not zero because the angle between r and v is never a right angle. Each current element in the long wire contributes a force to the charge q. The total force on q obtained by adding all these contributions is in the direction shown in Figure 9-10.

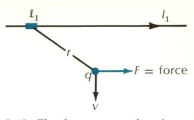

9-10 The force on q when its velocity is perpendicular to the wire.

To summarize, although the force between a moving charge and a current element is always an inverse square force as given by Equation (9-2), it is not always a **central force** (that is, it is not always directed along the line that joins the charge and the current element). The force may also have a component in the direction of the flow of current in the first current element.

Example 9-3:

What is the force on a negative charge of 10 ab-C moving at a velocity of 10^5 cm/sec directly toward a long wire carrying a negative current of 10^3 ab-A? The distance from the charge to the wire is 10 cm. Use $F = 2Iqv/r$, with $I = 10^3$ ab-A, $q = 10$ ab-C, $v = 10^5$ cm/sec, and $r = 10$ cm.

$$F = \frac{2 \times 10^3 \text{ ab-A} \times 10 \text{ ab-C} \times 10^5 \text{ cm/sec}}{10 \text{ cm}}$$

$$= 2 \times 10^8 \text{ dynes in the } \textit{opposite} \text{ direction to the current flow}$$

It is interesting to note that when the current in the long wire in Figure 9-9 produces an upward force on the charge moving away from it, the moving charge does not return an equal and opposite force to the wire. In fact, *it gives no force to the wire* because the wire is part of a complete circuit: The sum of the force contributions from current elements like e_1 due to the single moving charge cancels and yields no net force.

This violation of Newton's third law of action and reaction was not recognized until long after the Electromagnetic Revolution. While the reasons for this violation are beyond the scope of this book, we can state that, although Newton's third law does not always hold, the almost equivalent law of conservation of momentum does, even in this unusual case. Therefore, this example does not represent any new law or principle of electromagnetism, since it can be explained by a mathematical treatment of the basic electrodynamic equations we have presented here.

We have seen in this chapter that the force laws for electric currents (moving charges) are considerably more complicated than the simple inverse square force (Coulomb's law) for stationary charges or for stationary masses (Newton's law). In the following chapter, we will find that even more complications arise when we discuss magnetism, its relation to electric currents, and the subject of **electromagnetic induction**. Are the phenomena of nature merely becoming more complicated? Or could we be looking at electric phenomena in the wrong way? Are we faced here with the same situation earlier scientists experienced when they employed epicycles to define a planet's orbit only because they could not believe, as Kepler did, that a planetary orbit could be simple? We will see in Chapter 15 that this was the case—a fact unrecognized in physics until long after the Electromagnetic Revolution had ended. Electromagnetic theory could not be simplified until Albert Einstein proposed his "Special Theory of Relativity" in 1905.

Questions

1   A long, straight wire carries a current. What is the *direction* of the force this current exerts on a positive charge moving parallel to the wire and in the same direction as the current?

2   Ampère showed that the force between currents in two wires is a direct inverse square force similar to that between two charges. Yet Oersted's discovery showed that the force between a current and a magnet produces a circular or rotary motion. How can both of these phenomena be correct?

3   Place a clean copper penny on top of your tongue and a clean silver coin underneath it. Do you experience a peculiar taste? What apparatus have you just made?

4   Is lightning a natural source of electrical current? Explain.

5   Do you think the term "current" (as in a river current) is an appropriate way to explain the motion of electrical charges? Explain.

Questions Requiring Outside Reading

1   What were some of the early discoveries related to electricity and magnetism and what practical uses, if any, were made of these discoveries?

2   Oersted's discovery that electric currents produce magnetism was "accidental" in that he had no idea he would arrive at his end result and, indeed, was attempting to demonstrate the opposite principle. What other accidental discoveries have increased man's knowledge or understanding?

3   What controversy existed between Volta and Galvani concerning the production of electric currents? What similar scientific controversies can you describe?

1 Two long, straight wires carry currents in the same direction; the force attracting each one to the other is 5 dynes. If the distance between the wires is doubled, what is their attracting force?

2 Two long, straight, parallel wires A and B (see Figure 9-11) carry equal currents I_A and I_B in opposite directions. A third wire C, parallel to the other two, carries a current I_C which flows in the same direction as I_A. What is the direction of the force on wire C?

3 A magnet is placed next to a current-carrying wire and is allowed to turn until its preferred orientation is reached. What happens to the magnet if the current is reversed? Draw a diagram.

4 An electron is moving parallel to a current-carrying wire; the force acting on the electron is 4×10^{-2} dynes. What will the force be if the current in the wire is cut in half?

5 What would the force in Problem 4 be if the size of the electron's charge were twice as great?

6 A light, loosely-coiled spring (shown in Figure 9-12) is connected in an electrical circuit. When a current I flows, will the spring expand or contract? What causes this phenomenon?

Problems

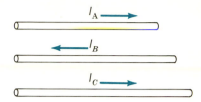

9-11 Arrangement of currents in Problem 2.

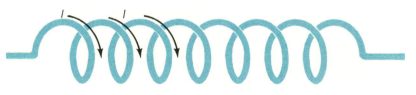

9-12 The coiled wire in Problem 6.

7* Two long wires, each with a mass m g/cm, are carrying currents in opposite directions. They repel one another. If one wire is placed directly above the other, at what distance d will it remain stationary if each wire carries a current I? (Your answer should be given in terms of m, g, and I.)

*If you omitted Section 9-5, disregard Problem 7.

10

Magnetism and

Electromagnetic Induction

10-1

Introduction

The current-current forces on a moving charge or on a current-carrying wire are traditionally explained in terms of **magnetic** fields. Instead of using Ampère's current-current force [Equation (9-2)], the magnetic field is introduced as an intermediary between the two wires or the two current elements. Although more conventional, the magnetic field method is also more complicated because of the essential three-dimensional description of the interaction of currents and magnetic fields. As we will see, the force between two long, parallel wires is not arrived at in as simple a way by the magnetic field method as it was in Chapter 9 (pages 124-26). When the magnetic field is introduced as an intermediary, there are two rules: First, one obtains the magnetic field due to the first current I_1; secondly, one finds the force on the second current I_2 arising from the interaction of the current I_2 in the presence of the magnetic field (set up by I_1).

10-2

Electric Current Forces via the Magnetic Field

The first rule states that the field lies in a plane perpendicular to the wire, in the form of continuous **concentric circles** (see Figure 10-1). If the wire is long and straight, the strength of the field falls off inversely with the distance away from the wire. The field exists not only in the single plane illustrated here, but also in all parallel planes and down the wire.

To find the force on a second wire carrying a current I_2, the second rule tells us to multiply the value of the magnetic field at the position of the second wire by the current I_2 in that wire. However, the direction of the force is perpendicular *both* to the magnetic field and to I_2. Thus, a current directed north placed in a magnetic field directed west will experience a force straight *up!* The force between two currents I_1 and I_2 (obtained earlier from Ampère's relatively simple inverse square law) can now be arrived at via the magnetic field construct from Figure 10-2. According to the first rule, wire (2) is in a magnetic field set up by I_1 (shown by the concentric circles). The second rule gives the force direction as perpendicular to both I_1 and the field; thus, we find the force on I_2 in a direction toward I_1 to be in agreement with the inverse square result of Ampère's law. (We have only considered the magnetic field in a qualitative way here in order to explain Ampère's experiment on the attraction or repulsion between two parallel wires carrying currents.)

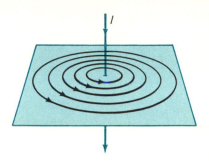

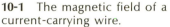

10-1 The magnetic field of a current-carrying wire.

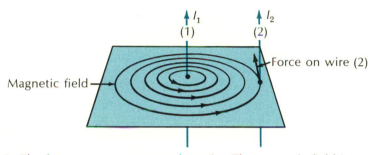

10-2 The force on a current-carrying wire. The magnetic field is set up by the current in wire (1).

We have seen how Ampère's inverse square law gives a simpler explanation of current-current forces than when the magnetic field is used as an intermediary agent. Next consider the forces of the **magnet-magnet** and the **current-current** interactions. These can also be explained by Ampère's law without using magnetic fields. Historically, Ampère postulated a magnetic theory in which permanent circular currents were produced by the **microscopic** constituents of magnets, so that all magnetic forces could be explained by the current-current inverse square law. His theory did not find acceptance, partly because it was almost 100 years before the atomic and molecular origin of his circulating currents were known and partly due to the tremendous success of the Faraday-Maxwell picture of electromagnetism which, as we will see, gives equal importance to both the electric and the magnetic fields. Taken as basic physical realities, the electric and magnetic fields supplant the Newtonian concept of "action at a distance."

Our theory of magnetism today is similar to Ampère's. If we look at a permanent magnet, we find that essentially all of the atoms in the magnet align so that circulating atomic electrons

10-3

Magnets

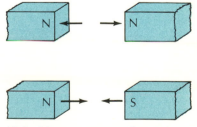

10-3 Magnet-magnet forces: Repulsion between like poles and attraction between unlike poles.

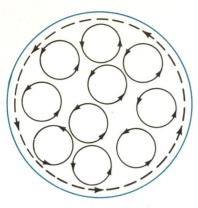

10-4 Circulating microscopic currents align in a permanent magnet.

rotate in the same circular direction in the atom (as in Figure 10-4). In some cases, the electron's intrinsic angular momentum, called **spin**, aligns and this also produces an equivalent circulating current. In the interior region, the circulating **microscopic**, atomic currents cancel each other, so that the combined effect is a large, **macroscopic** circular current (the counterclockwise dashed line in Figure 10-4). The difference between materials which can be magnetized and nonmagnetic materials lies in the ability of a material's atoms to align themselves in the presence of an externally applied magnetic field and to retain this orientation. Most nonmagnetic materials [gases, liquids, nonmetals, and nonferrous (nonmagnetic) metals] have randomly oriented atoms, so that the average magnetic effects of the many circulating atomic currents cancel. Magnetization represents a cooperative phenomenon among atoms; various materials or elements behave differently due to the internal structure and arrangement of their atoms. The subject of magnetism on an atomic level forms one branch of solid state physics and is being actively studied today.

10-4

Magnet-Magnet and Current-Magnet Forces

How do two magnets attract or repel one another? The magnetic field picture uses magnetic lines of force between a north and south pole (see Figure 10-5), much the same as the electric field picture uses lines of force between a positive and negative charge (see Chapter 8, Sections 8-4 and 8-5). The explanation using Ampère's current-current force is the one we will adopt here. Let us assume conventionally that a clockwise macroscopic current is a north magnetic pole when viewed on end and that a counterclockwise macroscopic current is a south magnetic pole (Figure 10-6). When a north pole is brought face to face with a south pole, the two circular currents are in the *same* direction and Ampère's current-current law predicts attraction (Figure 10-7). When two north poles or two south poles are brought together they yield respective circular currents in opposite directions and Ampère's law predicts repulsion (Figure 10-8).

The force between a magnet and a wire carrying a current can be explained in the same manner. In Figure 10-9, a vertical wire

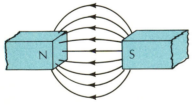

10-5 Magnetic "lines of force."

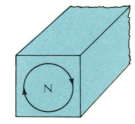 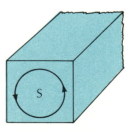

10-6 Aligned, circulating currents for north and south magnetic poles.

is placed between the poles of a horseshoe magnet and a current I_1 is directed down in the wire. Assume the wire is equidistant from the north and south poles (although this restriction is not necessary). The current-current force on the wire due to the circular current of the north pole is equal and opposite to the force on the wire due to the circular current of the south pole. This is evident since the direction of the circular current at the inner surface of the magnet is upward at both the north and south poles. The same is true for the further surface of the magnet where each pole has a downward circular current which affects the wire with an equal, but opposite, force. As a consequence, there is no net force on the wire from the north or south poles.

However, a magnet is magnetized by these circular currents throughout its entire volume. Let us consider the force on the wire inside the horseshoe magnet from the circulating current in the middle section of the magnet, halfway between the north and south poles (see Figure 10-10). Look at this midsection and the wire

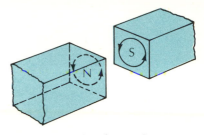

10-7 Attractive force: Currents moving in the same direction.

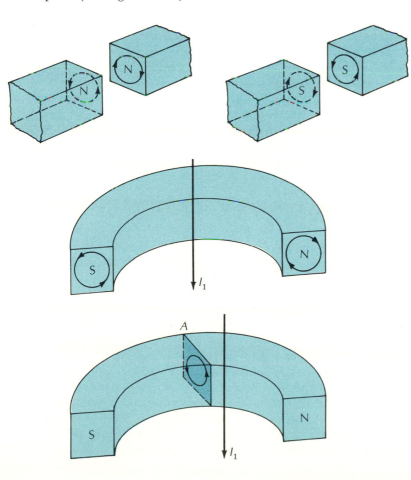

10-8 Repulsive forces: Currents moving in the opposite direction.

10-9 Wire carrying a current inside a horseshoe magnet.

10-10 Wire interacting with circulating current loop in cross-sectional area A of horseshoe magnet.

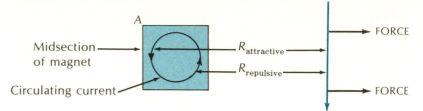

10-11 Close-up of wire and area *A* in the same plane.

in the same plane (Figure 10-11). The circulating current is counterclockwise. On the right-hand side, where it is nearest the long wire, the circulating current in the magnet is moving in the opposite direction from the current in the wire. By Ampère's law, there will be a repulsion dependent upon the square of the distance $R_{repulsive}$. On the left-hand side, the magnet current is flowing in the same direction as the wire current. There is an attraction dependent upon the square of that distance $R_{attractive}$. But, as we can see from Figure 10-11, $R_{attractive}$ is larger than $R_{repulsive}$ and the inverse square law predicts a net repulsive force away from the midsection of the magnet. If the current is reversed, the wire will be attracted, moving in between the poles and clinging to the midsection of the magnet.

From now on, we will use the current-current picture of Ampère's law. Magnetism and magnetic fields will only be discussed in an illustrative context to indicate their important historical roles in bringing about the next synthesis—Maxwell's Electromagnetic Revolution.

10-5

Michael Faraday and Electromagnetic Induction

Michael Faraday is one of the most unusual personalities in the history of physics. An anomaly among physicists, he entered into the study of the complex field of electricity without any formal training. Serving as laboratory assistant to Sir Humphrey Davy, a famous chemist in the early 1800's, young Faraday tried experiments of his own. A year after Ampère's and Oersted's individual discoveries of the magnet-current force, Faraday invented what was probably the world's first electric motor by utilizing the current-current force to produce a circular rotary motion. His discovery was the beginning of a whole new electrical technology.

Although he contributed greatly to chemistry and physics and to technology in general. Faraday's discovery of **electromagnetic induction** was his greatest and most important single achievement. He questioned that if a magnet had the ability to cause a force on a wire carrying electricity (Ampère's and Oersted's experiments), could the process be inverted so that a magnet could produce an electrical current in a wire. After many unsuccessful attempts, he found the answer in just as startling and unexpected a way as Oersted had made his earlier discovery of the rotary motion of the compass needle. Faraday found that indeed he could induce a current to flow in a wire when a magnetic field was present, but *only when the magnetic field was changing with time.* Thus, for example, a long, straight wire carrying a current could induce a current in a nearby wire, but only during the short time when the current in the first wire was either being turned on or turned off. Likewise, a current could be induced in a wire if the wire were moved in a magnetic field (that is, moved in the vicinity of a magnet) or if a magnet were moved near the stationary wire. But no induction of current occurred without either the *relative motion of a magnet and a wire* or a *changing electric current.*

It is interesting to note that Joseph Henry, a high-school teacher at the Albany Academy in New York, discovered the same phenomenon of electromagnetic induction in 1830, a year before Faraday did. However, due to a heavy eleven-month teaching schedule, he postponed publication of his work for a year and was unable to claim priority for the discovery. Following Benjamin Franklin, Joseph Henry is the second most important American physicist; he contributed greatly both to a basic knowledge in electromagnetic studies and to the technology of the subject.

There are two physically different phenomena that can induce a current in a wire (or induce an electric field in the region of space surrounding a wire and cause the current to flow in the wire—an equivalent statement). The *first* method of induction is the relative motion of the magnet and the wire. We will not use the magnetic field concept in our description here, but will consider induction to be caused by the motion of a wire relative to circulating current loops in a magnet (see Figure 10-12). The magnet is really composed of circulating currents and we can generalize that this first method of induction is due to the relative motion of a wire conductor with respect to another wire already carrying a current. Consider the two long, parallel wires in Figure 10-13 with a current I_1 flowing through one wire. When the second wire is moved away from the first (or vice versa), a current is induced in the second wire in the same direction as I_1. If the second wire is moved toward the first, the induced current will be in the opposite direction.

We can understand induction due to the relative motion of two wires simply by using Ampère's law for the force on moving charges. Consider the current in the first wire moving upward as shown in Figure 10-13. When the second wire is moved to the right as indicated by the arrows in the second part of the figure, the

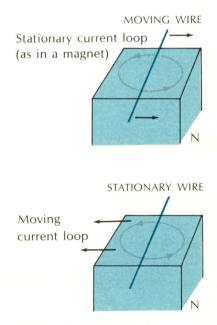

10-12 The first kind of electromagnetic induction: relative motion of a wire and a magnet.

Michael Faraday with his wife.

Radio Times Hulton Picture Library

electrons in the metal of the second wire are confined to the wire and move to the right as the wire is moved. But we have already learned from Ampère's law that there is a force on an electron moved at a velocity $\vec{v}$ away from a wire carrying a current, and that this force is in the same direction as the first current. Thus, all the electrons in the second wire feel this upward force and commence to move upward. This **induced force** is equivalent to

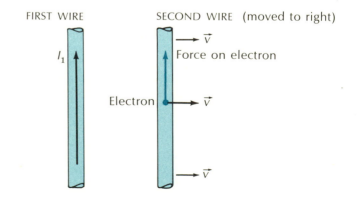

FIRST WIRE   SECOND WIRE (moved to right)

I_1

$\vec{v}$
Force on electron

Electron $\vec{v}$

$\vec{v}$

10-13 The first method of electromagnetic induction obtained from Ampère's law for the force on moving charges.

FARADAY

Michael Faraday was born to a London blacksmith in 1791. He was greatly influenced by his father, a member of the Sandemanians—a fanatically religious group who acknowledged the Bible as the source of all truth and who followed Apostolic practices. The sect intermarried, did not proselytize, and felt that saving money was sinful. Faraday himself later became a Sandemanian and eventually was elected to the position of elder. Elders were expected to be at meetings every Sunday and once when Faraday, although by then internationally famous, missed a meeting because he was dining with the queen, he was suspended. Throughout life, Faraday was to maintain his deep religious ties. He considered religion to be a higher order of truth than natural truth and, believing that religion contained the answers to all questions of human destiny, he never wasted any time on philosophy. □ At the early age of 13, Faraday went to work for a bookseller. There he learned bookbinding and eventually became interested in reading the end results of his labor. His broadened interests incorporated science, and he attended a series of lectures given by Sir Humphrey Davy—a leading chemist in England. When his apprenticeship as a bookbinder was over, Faraday wrote Davy, asking him for a job and including his carefully written and handsomely bound notes from Davy's lectures. His scientific career had begun. □ When Davy decided to tour the science centers of Europe, he asked Faraday to accompany him as his assistant and valet. Never having been more than 12 miles from London in his life, meeting the most important scientists of the age did a great deal to enlarge Faraday's narrow view of the world. □ Faraday did not write his first paper until he was 25, and then it was only a 400 word treatise on the analysis of some caustic lime. He did not publish any important discovery until he was 29, making him one of the slowest-blooming geniuses in modern science. □ One of his early discoveries, based on Oersted's work, resulted in the first electric motor. Unfortunately, Faraday's old mentor Davy had worked on a similar device earlier and considered Faraday's work to be plagiarism. Although Faraday was able to convince Davy's colleague in the latter's earlier experiments that he had arrived at his discovery independently, Davy himself opposed Faraday's admission into the Royal Society. Friction between the two continued, although Davy finally recommended Faraday's appointment to Laboratory Director of the Royal Institute. □ Faraday seemed to have enjoyed himself immensely even though his social life was very limited. Among his activities, according to one biographer, were bicycle races he used to conduct around the Institute with his cousins. □ Faraday's most important discovery was the principle of electromagnetic induction in 1831, which led to the development of the electric generator. Around 1840, Faraday had a mental breakdown mainly attributed to overwork (at 49, the same age at which Newton's breakdown occurred). After a long vacation, Faraday returned to his research, but refused the presidency of the Royal Society in 1857, partly because he disagreed with the society's constitution and partly for reasons of health. He died in 1867, at 75.

an electric field in the second wire and gives rise to the induced electric current in that wire.

The *second method of induction* cannot be deduced from Ampère's law; instead it must be described in terms of changes in currents. In the example of two parallel wires shown in Figure 10-14, assume that each wire is fixed and has no current. If one wire is then connected to a battery, as a current I_1 begins to

10-14 Second method of electromagnetic induction described in terms of current changes.

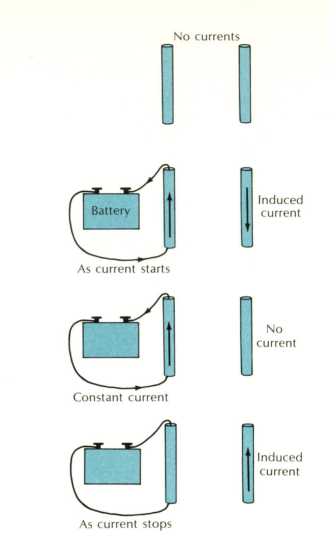

No currents

Battery

As current starts

Induced current

Constant current

No current

As current stops

Induced current

flow in the first wire, a current I_2 flowing in the opposite direction to I_1 is induced in the second wire. Once the original current I_1 in the first wire becomes constant and steady, there is no induced current in the second wire. When the first wire is disconnected, however, we again find an induced current in the second wire as the current I_1 drops to zero, but this time *in the same direction as I_1*.

Both phenomena of induction can be accounted for in a qualitative way. If the current elements in the first wire $e_1 = I_1 \ell_1$ or $e_1 = qv_1$ (where v_1 is the drift velocity of electrons of the first wire) are changed relative to the second wire, then a current is induced in the second wire. By stating that induction depends upon changing the current element e_1, we have not been very precise since we have not specified the geometry or relative orientation of the two wires. However, changes in $e_1 = I_1 \ell_1$ qualitatively account for the first method of induction changing relative motion (changing ℓ_1) as well as for the second method of induction changing current (changing I_1).

We will leave the subject of electromagnetic induction for now, keeping in mind that (1) Coulomb's law, (2) Ampère's law, and (3) Faraday's law of induction outlined above constitute the basic laws of electromagnetism in language which eliminates the magnetic field construct. To fully understand the great synthesis that came after Faraday, we must now examine an entirely separate branch of physics—the study of light.

Questions

1 Are there any philosophical problems associated with electricity and magnetism as profound as those associated with the earlier revolution in physics influenced by Newton, Galileo, Bruno, Kepler, and Copernicus?

2 Do you think the medieval church which arrested Galileo and burned Bruno at the stake would have been as rigidly opposed to the propagation of the work in electricity and magnetism of the early 1800's? Would any authority other than the church have opposed it?

3 Explain why magnetism is considered part of the study of electricity. Why aren't the two treated separately?

4 Can you explain what is meant by the statement that the electrical force is the most important force in nature?

5 Compare the idea of "action at a distance" with that of "force fields" such as electric and magnetic fields filling all of space. Is one idea more plausible than the other?

6 Are electric and magnetic fields real or are they merely calculative or interpretive devices? Explain.

7 In what sense are the physical phenomena which we describe *real* and in what sense are they simply devices for relating the realities of a physical situation to our own peculiar ways of perceiving things?

8 Faraday lacked any formal scientific training, yet he was a brilliant physicist and contributed greatly to the understanding of electricity and magnetism. Do you think he was hampered or helped by his lack of formal training, especially in the field of mathematics?

9 Faraday attempted to use magnets to make currents flow in wires. He reasoned that if currents produce magnetism, then magnetism must produce currents—that nature must be symmetric in this respect. Can this idea of "symmetry" be applied to natural phenomena in this way?

10 Why does adding iron inside a current loop concentrate and increase the magnetic field lines? Why is the magnetic field produced by a coil wound around an iron far much stronger than the field produced by a coil wound around wood or plastic?

11 Where is the source of the magnetic field of a permanent magnet? Compare a magnetized piece of iron with an unmagnetized one.

12 Does a magnetic field exert a force on a stationary electron? Explain.

13 Is the direction of the magnetic force on a moving electron parallel to the velocity of the electron? Is it parallel to the magnetic field? Explain your answers.

14 What is Faraday's law of induction? What uses are made of it in our technology today?

15 Is the magnetic field between two parallel wires carrying the same current stronger or weaker than the field at the same point when only one wire is carrying current? Does the direction of the currents matter? Explain.

Questions Requiring Outside Reading

1 Explain the "Right-hand Rule" for producing magnetic fields from currents.

2 Explain the "Right-hand Rule" for producing a force on a wire carrying a current in a magnetic field.

3 What are magnetic poles? Can a single north or south magnetic pole ever be found in nature? Explain.

4 What other contributions has Faraday made to science in addition to those discussed in this chapter?

5 What controversy existed between Davy and Faraday?

Problems

1 Three long, straight wires A, B, and C pass through the vertices of an equilateral triangle (see Figure 10-15). Each carries the same amount of current in the same direction. Sketch the magnetic field lines around the three wires.

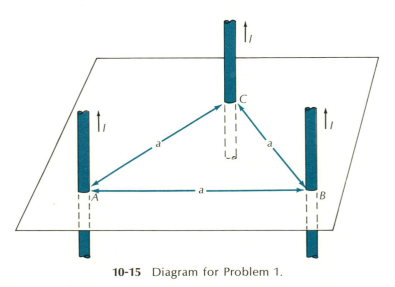

10-15 Diagram for Problem 1.

2 What is the direction of the force on wire C in Problem 1?

3 A long, straight wire carries a current I. If a second wire which initially carries no current is moved toward and parallel to the first, in which direction will the induced current in the second wire flow?

4 Describe how you might use a permanent bar magnet and a coil of wire to generate electric currents. Why would this current alternate in direction? Draw a diagram.

5 Explain why the strength of a "permanent" magnet might be reduced by heating it.

6 Use Ampère's law to explain why two moving electrons repel each other to a lesser degree than two stationary electrons. If both electrons are moving parallel to one another with a speed v, how much weaker is the force between them than if they were stationary?

7 Use the arguments in Section 10.4 (page 134) to explain why two bar magnets placed beside one another exhibit repulsive or attractive forces.

11 Light

11-1
Introduction

Both the origin and mechanism of light have been speculated upon by philosophers and scientists from the beginning of recorded time. The early Greeks and later astronomers like Hipparchus and Ptolemy wrote treatises on light and optics. Some believed that vision resulted from particles sent out by the human eye; others, including Archimedes and the Pythagoreans, realized that the ability to see was determined by the source of light (the emission theory). The latter group also knew much about **geometrical optics** (the treatment of light as rays that travel in straight lines). Light rays are sometimes bent—by a mirror (reflection), by water (a stick appears to bend when half of it is immersed in water), by a glass prism. During the middle of the seventeenth century, the inventions of the telescope, the microscope, and other optical instruments using lenses began to interest scientists in studying the principles of optics. Newton, Kepler, Huygens, and Descartes were major contributors to such research.

11-2
Is Light a Particle or a Wave?

In addition to the *Principia,* Newton's other major treatise *Optics* contained the results of many of his experiments and theories regarding light. Newton showed that a glass prism could bend a narrow ray of white sunlight and break it up into the colors of the rainbow. (This is called the **dispersion** of light; see Figure 11-1.)

Newton concluded that white light must be composed of all the colors in the spectrum—red through violet. Since he believed that the various colors were bent to different degrees when **refracted** by the glass prism due to forces that changed their direction of motion, it was natural for Newton to consider light to be composed of particles which obeyed the force laws set down

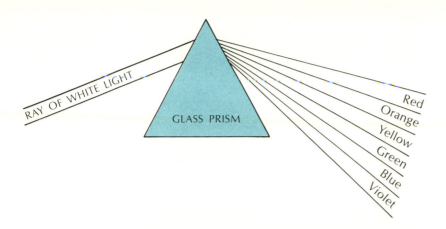

11-1 White light broken up by a glass prism into the spectrum of colors.

in the *Principia*. Particles of violet light would have less mass than particles of red light, since the violet light with a smaller inertial mass is bent more. Furthermore, the facts that illuminated objects cast sharp shadows and that geometrical ray optics produces accurate results lend more support to this particle theory of light.

However, Christian Huygens, a contemporary of Newton, worked out a different theory of light based on waves instead of particles. Huygens believed that light was emitted in waves from a source, in much the same way as waves are produced when a stone is dropped into a pool of calm water. From Newton's and Huygen's time to the early 1800's, the controversy over the nature of light remained unsettled; Newton's particle theory was the more generally accepted of the two, probably due to the successes of his work in gravitation. Around 1800, however, a physician, Thomas Young, performed a definitive experiment which showed that light *must behave* like a wave. Light from a point source was intercepted by an opaque obstruction in which two very narrow, parallel slits were positioned quite close together. Two light beams emerged from the other side of the obstruction, one from each slit. If light consisted of particles, he felt, then the geometrical image or shadow of the two slits could be seen on a screen placed behind the obstruction; the screen would be dark except for the illumination by the two slits S_1 and S_2 (see Figure 11-2). But this

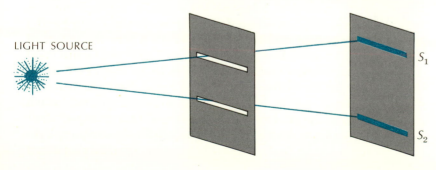

LIGHT SOURCE

11-2 Illumination of two narrow slits: Newton's particle theory.

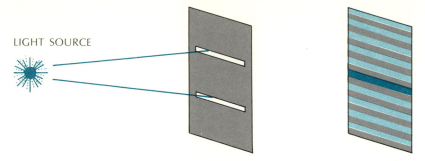

LIGHT SOURCE

11-3 Illumination of two narrow slits: Young's actual experimental results.

is not what Young observed. When the distance between the slits was very small, he found that only *one* bright slit image midway between S_1 and S_2 appeared on the screen while above and below this bright slit were many alternating dark and bright slit images (Figure 11-3; see Figure 11-2 for positions of S_1 and S_2). Clearly this result could not be explained by Newton's particle theory of light, but it could be completely explained by Huygen's wave theory. To understand why this is so here, we must first look at some of the simple properties of waves.

11-3

Waves

The waves generated when a stone is dropped into water or produced when we hear sounds have four features in common:

1   They are a disturbance of a **medium** (water molecules for water waves; air molecules for sound waves) and the disturbance moves in that medium with a **definite velocity v**.

2   The waves have an **amplitude** A which measures the amount of the disturbance. The magnitude of the crests and troughs in a water wave clearly illustrate the amplitude (see Figure 11-4).

3   The distance between two successive crests or troughs is always the same for a wave traveling in a homogeneous medium (of same composition everywhere). This distance which is called a **wavelength** is represented by the Greek symbol λ (again, see Figure 11-4).

4   As the wave moves by us, the number of crests that travel past any given point per second is the **frequency** f of the wave (that is, the number of vibrations per second).

If we watch a particular crest of a wave, we will see it traveling along with a constant velocity v. In a given time t, it will have traveled a distance of $s = vt$. In this distance s, there are n wave-

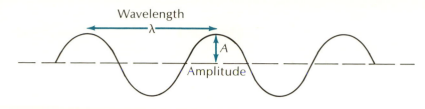

Wavelength
λ

A
Amplitude

lengths λ, where $n = s/\lambda = vt/\lambda$. But n is also the number of wave oscillations (vibrations) in the time t. If the wave vibrates at f oscillations per second, then ft equals the number of oscillations n that go by us in the time t. Equating these two values of $n = vt/\lambda$ and $n = ft$, we find the relation of v, λ, and f:

$$n = \frac{vt}{\lambda} = ft$$

Canceling t on both sides, we obtain the basic equation for any wave,

$$\frac{v}{\lambda} = f \qquad \text{or} \qquad v = f\lambda \qquad \text{(11-1)}$$

Example 11-1:

A sound wave travels in air with a velocity of about 400 m/sec. If the frequency of "middle C" on a piano is approximately 250 vibrations/sec or 250 sec^{-1}, what is the wavelength of "middle C?" Use $v = \lambda f$ or $\lambda = v/f$, with $v = 400$ m/sec $= 4 \times 10^4$ cm/sec and $f = 250$ sec^{-1}. Then

$$\lambda = \frac{4 \times 10^4 \text{ cm/sec}}{250 \text{ sec}^{-1}} = 160 \text{ cm}$$

A periodic wave with crests and troughs occurring at regular intervals (as shown in Figure 11-4) is the solution of what is called a differential equation. This equation, normally only written with the calculus, will be presented here in words without calculus. First, consider the **oscillatory** motion of a weight on the end of a spring. This motion is **periodic** in time, since it retraces its path over and over with a constant frequency (the number of periodic oscillations per second). A pendulum also swings back and forth with a periodic motion. The physical feature common to all periodic motion is that the distance of the oscillating object from the midpoint of the motion is always proportional to the acceleration (or force) trying to reverse the motion so that the object will oscillate in the opposite direction. Thus, when the spring is most elongated and the weight is furthest from the midpoint of the motion, the upward force of the stretched spring is the largest.

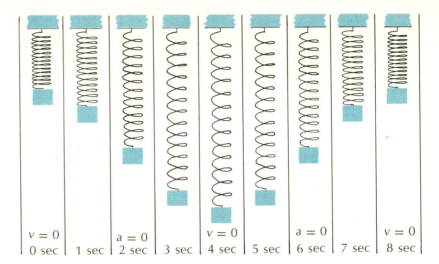

$v = 0$		$a = 0$		$v = 0$		$a = 0$		$v = 0$
0 sec	1 sec	2 sec	3 sec	4 sec	5 sec	6 sec	7 sec	8 sec

11-5 Periodic motion of a spring pendulum.

The periodic motion of a spring pendulum with a time period of 8 sec is shown in Figure 11-5. An equation for this periodic motion can be written as:

$$a = -ks \tag{11-2}$$

This equation states that the acceleration a which tries to change the motion of an object is proportional to its **displacement$\vec{s}$** from **equilibrium**, but in the opposite direction. The symbol k is a constant which enables the proportional statement to be made into an equation. Equation (11-2) holds for all "simple" periodic motion. (Other more complicated types of periodic motion—the elliptical periodic motion of planets about the sun, for example—are described by more complicated equations.)

But what does all of this have to do with waves? Waves have a periodic motion in time as well as a periodic motion in distance (along the direction of the velocity). This can be described in words as: Periodic change of motion with time = periodic change of motion with distance. A wave's oscillation with time is equal to its oscillation with distance as it travels along. The exact relation is given by Equation (11-1) which states that the number of oscillations in time $n = ft$ must equal the number of oscillations over the distance traveled $n = vt/\lambda$.

It is important to ask here what causes an oscillatory motion in time to travel outward in space like a wave. In the case of a wave in a rope, the tension caused by a sudden displacement of the rope is transmitted to other nearby points along the rope due to the fact that tension is felt along the entire stretched rope. In Figure 11-6 we see that by shaking the rope up violently, we force the rope upward at point A. If we quickly bring the left-hand side

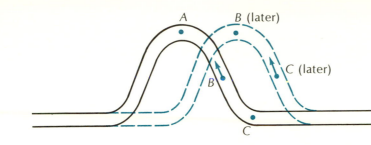

of the rope down again while it is still moving up, we can force point *A* to move to the right. The tension in the curved piece of rope shown in Figure 11-6 then produces an *upward force* on a nearby point *B*, so that *B* moves upward to the position shown within the dotted lines. At this position, a nearby point *C* feels the tension in the rope and experiences an upward force. Thus a wave is propagated because of local tensions in the vicinity of the disturbance which first caused the wave.

In a water wave, surface tension produces the local stresses that cause the wave to move outward when a stone is dropped into water. In a sound wave, the collisions of air molecules cause the wave to move away from the source with the velocity of sound. All these waves have one feature in common: They have an **elastic medium** (the rope, the water, the air molecules) which, by being compressed or stretched, causes the wave to **propagate**. The wave will move outward at a constant velocity that is dependent upon the elastic properties of the particular medium. The apparent conclusion from this discussion is that, from our experience, *waves require an elastic medium on which to travel.*

The energy that a wave transmits as it travels from one point to another must also be due to the stresses and tension of the medium. Instead of the kinetic energy of a projectile moving along a trajectory being equal to $\frac{1}{2}mv^2$, the wave's energy is derived from the potential energy of the disturbance (the tension or compression) of the medium. Since this disturbance propagates outward with wave velocity v, the energy also propagates outward. The energy can be obtained by starting with Equation (11-2). As we saw from Equation (11-2), in periodic motion the acceleration of the disturbance is proportional to the distance or displacement of the wave amplitude from the midpoint of the motion. But by Newton's second law, the force causing this disturbance is proportional to the acceleration; thus, the force is proportional to the wave amplitude. Now recall that energy is defined as Force × Distance or, in this case, Force × Amplitude. Consequently, the energy in a wave is proportional to the *square* of the amplitude of the wave

$$\text{Energy} = E \propto A^2 \qquad \textbf{(11-3)}$$

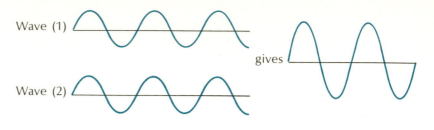

Wave (1)

Wave (2)

gives

11-7 Two waves added *in phase*.

11-4

Interference of Two Waves and Young's Experiment

What happens when two waves of the same amplitude are added together? If the two waves can arrive at some given point where each wave is at a crest or at a trough, then the two amplitudes combined will be exactly twice the amplitude of a single wave alone. At this point, the two waves are said to be **in phase**; note that the energy will then be four times the energy of a single wave alone (see Figure 11-7).

But if the two waves arrive at a given point where one is at a crest while the other is at a trough, the crest cancels the trough and the two waves give a combined amplitude of zero. The two waves are then **out of phase** and the energy is zero (Figure 11-8). If the two waves are neither completely out of phase nor completely in phase, an intermediate situation exists where the combined amplitude and the energy are each greater than zero, but less than their maximum values (see Figure 11-9).

Now the **interference** of the two light waves coming from the slits in Young's experiment (page 145) can be understood. Each slit is illuminated equally by the light and serves as a separate, but equal, light source, illuminating the screen with waves of equal amplitude which travel out in all directions from each slit. Furthermore, the light at the upper slit must be *in phase* with the light at the lower slit, since the distance of each slit from the source is assumed to be the same. The point A' on the screen (see Figure 11-10) is directly opposite the midpoint A between the two slits. Thus, the distance BA' which the light traverses from the upper slit B to A' is equal to the distance CA' which the light traverses from the lower slit C to A'. Since the waves were in phase at B and C, they must still be in phase at A'. Therefore,

Wave (1)

gives

Wave (2)

11-8 Two waves added *out of phase*.

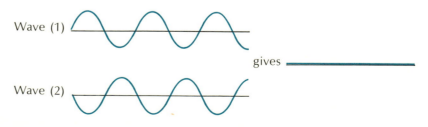

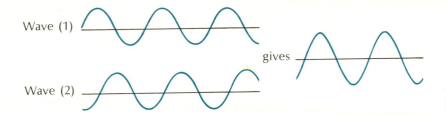

Wave (1)

Wave (2)

gives

11-9 Two waves added neither in nor out of phase.

the spot A' on the screen will be illuminated and will have a bright slit image.

We can find another spot D on the screen where the path distance CD is exactly one-half wavelength longer than the path distance BD (again, see Figure 11-10). Since the waves were in phase at B and C, they will now be out of phase at D and the spot D will be dark on the screen. Beyond D, there will be another bright spot, corresponding to a path difference from C of one whole wavelength over the path from B. Beyond that, there will be another dark spot, then another bright one, and so on. The light and dark spots will be observable if the slit separation BC is not too much larger than the wavelength of the light.

One last feature of light important in the electromagnetic synthesis is that the vibrations of the light wave are found experimentally to be a **transverse wave**. The electric field is in the plane perpendicular to the direction in which the light moves. (Water waves are also transverse, since the motion of the water particles is

11-5

Polarization of Light

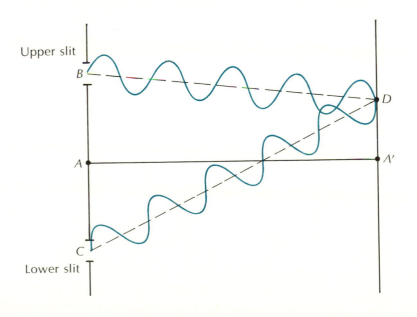

Upper slit

B

D

A

A'

C

Lower slit

11-10 Young's double-slit experiment.

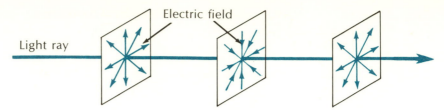

Electric field

Light ray

11-11 Vibrations of the electric field are perpendicular to the direction of the light ray.

vertical while the water wave travels out horizontally. Sound waves, on the other hand, are **longitudinal**—not transverse; the disturbance that produces sound is the motion of the air molecules colliding with one another in the *same* direction in which the sound wave moves.)

The vibrations that constitute ordinary light are in all possible directions in this perpendicular plane. In Figure 11-11 we have shown only three sample planes out of the infinite number that are perpendicular to a light ray. In any one of these planes the vibrations are periodic, back and forth away from the ray and toward the ray in all possible directions. (Ordinary light vibrations in all directions are analogous to all the possible orientations of a handful of toothpicks randomly dropped on the floor.)

Now there are certain materials (like Polaroid Sunglasses) which eliminate the vibrations or the disturbance of light waves in all directions except one. The light is then said to be **plane polarized**. The disturbance that causes the light to propagate is no longer over the entire plane perpendicular to the velocity, but is along one line or one direction within that plane. (This is analogous to dropping the randomly-oriented toothpicks over a grating (Figure 11-12) in that only those toothpicks that are parallel to the grating will pass through.)

In the case of a light ray, let's assume this direction is horizontal. Then if we place a second Polaroid in the path of the plane-polarized light, no light will be transmitted through the second polarizer if it allows only vertical disturbances to pass. In such a situation, the two polarizers are said to be crossed. It was found in the early 1800's that certain liquids like sugar solutions, when placed *between* two crossed polarizers, allowed some light to come through on the far right-hand side of the second polarizer. The reason for this is that these liquids have asymmetrical molecular structures which can *rotate* the plane of polarization from its horizontal direction to a more vertical one. As a result the second vertical polarizer allows a considerable amount of the light to pass through.

We are interested in this effect here because in 1847 Faraday

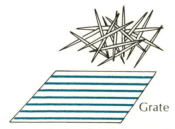

Grate

11-12 Randomly-oriented toothpicks passing through a grating.

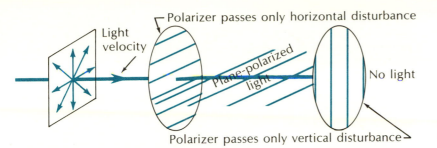

Light velocity

Polarizer passes only horizontal disturbance

Plane-polarized light

No light

Polarizer passes only vertical disturbance

11-13 Light passes through crossed polarizers.

found that a *magnetic field* would also affect light in this way. By allowing plane-polarized light to pass through a magnetic field set up between two crossed polarizers, Faraday observed that some light came through the second polarizer. In other words, the magnetic field had caused the plane of polarization to rotate—the first clue of any connection between electromagnetism and light.

11-6

Electric Units and the Velocity of Light

The second clue of such a connection was found about ten years later. In Germany, Wilhelm Weber made very accurate measurements of the ratio of the electrostatic system of units for charge Q (stat-C) to the electromagnetic system for charge q (ab-C). In Chapter 9, we saw that Q did not even have the same units as q, certainly most peculiar in itself, and that the ratio Q/q had the units of a velocity. Weber measured the total charge Q on a charged condenser (a Leyden jar). Of course this value of Q was in electrostatic units or stat-C. Weber then allowed this charge to flow through a galvanometer, which measures the amount of current in a wire by the amount of motion the wire exhibits when in the field of a strong magnet. He computed the force and found, from Ampère's law, that the current $I = q/t$ in electromagnetic units (ab-C) corresponded to the original charge Q. In this way Weber was able to measure the ratio as

$$\frac{Q}{q} = 3 \times 10^{10} \text{ cm/sec}$$

At almost the same time (around 1850), other physicists, notably Léon Foucault in France, had measured the velocity of light in air with great precision and had found that

Velocity of Light $= c = 3 \times 10^{10}$ cm/sec

Surely the equality of these two numbers was simply a coincidence. Or could there possibly be a connection between light and the electromagnetic laws of Coulomb, Ampère, and Faraday? The possible connection, certainly not an obvious one, was strikingly similar in its philosophical meaning to a connection made 200 years earlier when Newton had wondered if there were some relation between the gravitational force that caused accelerated motion *on* the earth and the force that kept the moon in its orbit *around* the earth. It took great imagination on his part to propose the inverse square force and to show that it indeed provided the required missing link between celestial and terrestrial motion.

With these few meager clues linking electromagnetism with light, James Clerk Maxwell was to show the same kind of creativity and imagination by providing the missing link between electromagnetism and optics in his proposal "A Dynamical Theory of the Electromagnetic Field" in 1864.

Questions

1 Can you apply the principle of superposition to white light? If so, in what way?

2 Do you believe that light travels in straight lines? Why?

3 What connotation do the terms *frequency* and *wavelength* have when related to waves?

4 Draw a diagram to show that the speed of a wave is given by $v = \lambda/T$, where v is the speed, λ the wavelength, and T the period of the wave.

5 Is the fact that the wave spreads beyond the slits important to the outcome of Young's double-slit interference experiment?

6 Look at a distant light source through a piece of finely woven cloth. What causes the strange appearance? Explain why.

7 How does the spacing of the slits affect the interference pattern in Young's double-slit experiment?

8 Explain why Young's experiment rules out Newton's particle theory of light.

Questions Requiring Outside Reading

1 Explain the formation of a rainbow. Of a double rainbow.

2 Compare the modern explanation of a rainbow with Seneca's explanation. With Newton's.

3 Explain how the astronomer Roemer unintentionally determined the approximate speed of light.

4 How do Polaroid sunglasses help to cut the glare from a road when driving?

Problems

1 Illustrate both the wavelength and amplitude of a wave.

2 Draw a diagram to explain how you can see your entire image in a mirror only half as tall as you.

3 Diagram constructive and destructive interference.

4 What relationship exists between wavelength, frequency, and velocity? Use this relationship to find the frequency of waves with a velocity of 500 ft/sec and a wavelength of 25 ft.

5 If waves with a frequency of 60 cycles/sec have a wavelength of 18 ft, what is their velocity?

6 If water waves move past you with a velocity of 5 ft/sec and the wavelength of these waves is 6 in., what is their frequency?

7 Draw diagrams to illustrate the interference of two different waves of the same wavelength and amplitude. Show how the two waves can be added together to give either no resulting wave or a wave with twice the amplitude of the original two. What principle is involved here?

8 Galileo and his assistant are standing on two hilltops 1.5 kilometers (approximately 1 mi) apart, trying to measure the speed of light by flashing signals back and forth with their respective lanterns. If each has a reaction time of 0.01 sec, show that they cannot measure the velocity of light c.

9 Two loudspeakers from a stereo set are placed 4 ft apart. You are standing 3 ft in front of one of the speakers, both of which are producing a tone of 512 cycles/sec. If the velocity of sound is 1024 ft/sec, can you hear the tone? Why or why not?

12

Maxwell and the
Electromagnetic Synthesis

12-1

Introduction

James Clerk Maxwell was born in Scotland in 1831, the year that Faraday discovered electromagnetic induction. Unlike Faraday, who dropped out of school at the age of 13, Maxwell had the benefit of a good education; he wrote his first scientific paper at 13, describing a mechanical method of drawing certain kinds of oval curves.

Maxwell exhibited the same powers of mathematical abstraction that Newton had, a quality which helped him to produce the synthesis of the Electromagnetic Revolution. However, Faraday did exert a significant influence on Maxwell in the sense that the magnetic and electric fields that Faraday used were also real entities for Maxwell. Maxwell used mathematics to simplify and codify the word-picture relations from Faraday's research. Through this mathematical abstraction of Faraday's observations, Maxwell was able to postulate a new physical "law" for which there had never been any direct experimental verification. This new law, coupled with the earlier empirical laws of Coulomb, Ampère, and Faraday, led to the prediction that light is a disturbance consisting of electric and magnetic fields propagating in space. Thus, the earlier clues began to make sense: We can begin to understand the rotation of the plane of polarization by a magnet if light is considered to be some kind of magnetic wave; the fact that the ratio of the electrostatic charge units to the electromagnetic units equals the speed of light *can* be more than a strange coincidence if light is considered to be an electric and magnetic field disturbance.

In his 1864 paper, Maxwell uses the electric and magnetic field concepts already introduced by Faraday. In Maxwell's words:

> I have preferred to seek an explanation in another direction, by supposing [the forces] to be produced by actions which go on in the surrounding medium as well as in the excited bodies [the charges and magnets], and endeavouring to explain the action between distant bodies without assuming the existence of forces capable of acting directly at sensible distances.

Note that Maxwell carefully avoids the Newtonian "action-at-a-distance" concepts.

Maxwell further argued the plausability of the existence of these electric and magnetic fields in the surrounding medium by appealing to the current knowledge of light and heat and the still generally accepted existence of the ether.

> We have therefore some reason to believe, from the phenomena of light and heat, that there is an aethereal medium filling space and permeating bodies, capable of being set in motion, and of transmitting that motion from one part to another, and of communicating that motion to gross matter.

Next Maxwell developed arguments for the plausability of a connection between electromagnetism and light by alluding to one of Faraday's earlier clues:

> Now we know that the luminiferous medium [the ether] is in certain cases acted on by magnetism; for Faraday discovered that when a plane-polarized ray traverses a transparent medium in the direction of the lines of magnetic force produced by magnets, the plane of polarization is caused to rotate. . . . We have therefore warrantable grounds for inquiring whether there may not be a motion of the aethereal medium going on whenever magnetic effects are observed.

Maxwell then explained his theory, presented equations, and showed that a wave equation for an electromagnetic disturbance was the natural consequence of that theory. The velocity of propagation of the electromagnetic wave predicted by his theory equaled the ratio Q/q of the electrostatic system of units to the electromagnetic system. He completed his synthesis with these words connecting the ratio Q/q with the speed of light c:

> By the electromagnetic experiments of Weber [see page 153], . . . 3.1×10^{10} cm/sec is the number of electrostatic units in one electromagnetic unit of electricity. The velocity of light in air is, according to the accurate experiments of Foucault, 2.98×10^{10} cm/sec.
>
> Hence the velocity of light deduced from experiment agrees sufficiently well with the value of Q/q deduced from the only set of experiments we as yet possess.

12-2

Maxwell's Synthesis

MAXWELL

James Clerk Maxwell was born in Edinburgh in June of 1831. His mother died when he was eight. His father, a lawyer more interested in keeping up with the progress being made in science than in spending a lot of time at his practice, raised his boy on a small farm. ☐ James was a very curious child. His earliest recollection was of looking up at the sun and "wondering" about it. He loved the country where he played imaginative games. (In one of these, he put frogs in his mouth and watched them jump out again.) ☐ Maxwell's early education left much to be desired. His tutor decided that James was slow to learn and, to speed things up, resorted to twisting his student's ears and similar tactics. James did not complain, however. He could be obstinate when he wanted to and had apparently decided to win the battle without the aid of outside intervention. ☐ In 1855, having gone through Cambridge, Maxwell published a major work on Faraday's lines of force. The paper, which contradicted widely-held electrical theories, was condemned by classical physicists. As if to prove to the classicists that he understood their kind of physics perfectly well, Maxwell examined Saturn's rings and showed that they were composed of many little satellites held together by the gravitational attraction of Saturn. An earlier antagonist in the lines-of-force debate called this discovery one of the most remarkable applications of mathematics to physics that he had ever seen. ☐ At this point, Maxwell decided to leave his position as a fellow at Cambridge and go to Aberdeen to be closer to his father who was very sick. His father died soon after, but Maxwell remained in Aberdeen for three more years, taking a position at the University and marrying the principal's daughter during his stay. ☐ After leaving Aberdeen, he applied for a teaching position at the University of Edinburgh, but it seems that he was not as good a teacher as he was a scientist and he did not get the post. In 1860, he landed a professorship at King's College in London, where he stayed for five years until he was forced to retire because of ill health. ☐ Maxwell maintained that his discoveries were not extremely difficult ones. Characteristically, he never lapsed into periods of introversion, but remained personable and unassuming even when in the midst of a theory. Maxwell was an excellent writer and had a facility for explaining things, uncommon among great scientists. He was always careful, however, not to let his factual models overburden his theories, so that he could abandon his mechanical explanations if necessary. ☐ In 1871, Cambridge University began work on a new scientific laboratory largely for the purpose of experimenting in heat and electricity, and Maxwell was persuaded to direct the project. He designed the building, the Cavendish Laboratory, which was finished in 1874 and which still operates today. Maxwell died in 1879, after a short illness.

The agreement of the results seems to show that light and magnetism are affections of the same substance, and that light is an electromagnetic disturbance propagated through the field according to electromagnetic laws.

This passage crowns Maxwell's synthesis. Note the striking parallel between his "agrees sufficiently well" (concerning the value of Q/q and the velocity of light) and Newton's "answer pretty nearly" (comparing the inverse-square gravitational force on the earth's surface with the force at the orbit of the moon; see Chapter 6, pages 72 and 77).

Newton and Maxwell both used clues provided by the comparison of quantitative measurement in one field of physical science with quantitative measurement in an *entirely different* and totally unrelated field. Newton made the synthesis between celestial and terrestrial phenomena by proposing a mathematical equation describing the inverse-square gravitational force; Maxwell made the synthesis between electromagnetism and light by proposing mathematical equations which will be described later in Section 12-4.

It is also interesting to note that Newton and Maxwell were both modest enough to give specific credit to their scientific predecessors. Newton said that if he had seen farther than others, it was because he had stood on the shoulders of giants, an obvious reference to Kepler, Galileo, and Copernicus. Maxwell, too, gave credit where credit was due:

> The conception of the propagation of transverse magnetic disturbances to the exclusion of normal ones is distinctly set forth by Professor Faraday in his Thoughts on Ray Vibrations (Philosophical Magazine, May 1846). The electromagnetic theory of light, as proposed by him, is the same in substance as that which I have begun to develop in this paper, except that in 1846 there were no data to calculate the velocity of propagation. [The ratio Q/q had not yet been measured.]

From what we have seen of Newton's creative powers and from what we are now seeing of Maxwell's genius, it is fair to say that each underplayed his own role in his respective revolution. Like Newton, Maxwell built upon the discoveries of his predecessors; however, Maxwell's use of mathematics surpassed the computations of any of the previous electromagnetic scientists. Maxwell, like Newton, was a man of peculiar genius whose creative imagination enabled him to see beyond any of his colleagues. His intuition led him to postulate a new force law—a new and as yet unobserved relation between the electric and magnetic fields. And, like Newton, he had the mathematical ability to use his intuitive insight to construct a completed theory of extraordinary scope.

12-3

Maxwell and Newton

The Granger Collection

James Clerk Maxwell

12-4

Maxwell's Equations in Terms of Magnetic Fields

We will now write the four equations that describe Coulomb's, Ampère's, and Faraday's laws as they appeared in terms of electric and magnetic fields *before* Maxwell's electromagnetic synthesis. The equations are not quite the same as those previously given since we are using the magnetic field construct here in order to distinguish Maxwell's relation between the electric and magnetic fields. The four field equations are:

1 The net number of electric field lines entering any volume of

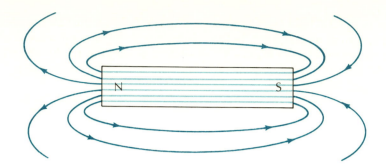

12-1 The magnetic field of a bar magnet.

space is equal to 4π times the net charge enclosed by the volume [see Equation (8-5), page 112].

2 The net number of magnetic field lines entering any volume of space is always zero.

3 The change in time of magnetic field lines sets up a circular electric field around the magnetic lines.

4 A steady current I sets up a circular magnetic field around the current.

These equations tell us how to find the electric field from its sources (either charges or a changing magnetic field) and how to find the magnetic field from its sources (currents). Once we find the electric and magnetic fields, we can find the motion of particles in these fields from the force laws. The force on a charged particle in a combined electric and magnetic field is

$$F = QE + Bqv \tag{12-1}$$

where v is the velocity of the particle, E is the electric field, and B is the magnetic field.

Maxwell's first law above is simply a restatement of Coulomb's law discussed earlier (page 112). If there are no free charges enclosed by the volume, then the net number of lines of force is zero (as many lines enter the volume as leave it). This first law follows from our assumption that the electric field lines begin and end on charges.

The second law is the magnetic field equivalent of the first law. It states that there are *no free magnetic poles* or magnetic "charges"; thus the magnetic field lines cannot begin or end, but must form continuous loops (see Section 10-2, page 132). In a bar magnet, the lines of force go through the magnet itself and out into space (see Figure 12-1).

The third law is a word statement of Faraday's theory of electromagnetic induction. If the magnetic field is changed with time (by moving the magnet or by starting and stopping a current), an electric field E is induced in a circular direction around the magnetic field (see Figure 12-2). Remember this only happens as

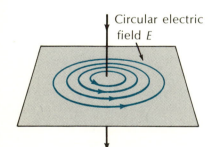

Circular electric field E

Increasing magnetic field B

12-2 An electric field is induced by changing the magnetic field.

the magnetic field is changing; an electric field cannot be induced by a *constant* magnetic field (see Figure 12-3). If a circular wire were present (as in Figure 12-4), electrons moving in the electric field would produce an electric current while the magnetic field was changing.

Maxwell's fourth equation gives the circular magnetic field set up by the current I (which was described in Section 10-2, page 133). Equation (12-1) relates the observed force to these fields. The first term QE is the force on a charge Q due to the electric field E. This force law is equivalent to Coulomb's law for point charges. The second term of Equation (12-1) gives the force on a moving charge due to the magnetic field B. When this force law is coupled with Maxwell's fourth equation (which states that magnetic fields are due to currents I), we have a result equivalent to that for the force on a moving charge due to a current in a wire (as discussed in Section 9-4, pages 126–29, where the magnetic field construct was bypassed).

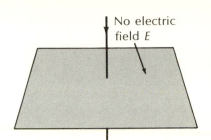

No electric field E

Steady, unchanging magnetic field B

12-3 No electric field is induced by a constant magnetic field.

You may wonder why we have treated Maxwell's equations in such detail and, particularly, why we have used the magnetic field concept, since previously we wrote the equivalent equations in a way that did not require the magnetic field. Our reason for repeating the electromagnetic equations in this form is so that we can understand the importance of the contribution Maxwell made when he added a term to these equations. Perhaps influenced by Faraday, he had already been thinking that a magnetic disturbance propagated like a wave in space. If we look at the third equation in Section 12-4 again, we see that the change of magnetic field B with time sets up a circular electric field E around itself. Now *if* a changing electric field E could set up a circular magnetic field B around itself, then the electric and magnetic field phenomena would be completely **symmetric** with respect to one another. Furthermore, we would have the basis for **electromagnetic wave** propagation in the ether: A periodically-changing magnetic field could set up a circular, periodically-changing electric field which could, in turn, set up another circular, periodically-changing magnetic field, and so on. Thus the waves could propagate outward in space as periodically-changing electric and magnetic field disturbances.

But nowhere in the four equations do we see that a changing electric field sets up a circular magnetic field. The fourth equation states that only a steady current I sets up a circular magnetic field, which does not help us because there are no actual steady electric currents (as there are in wires) when light travels in the ether.

12-5

Maxwell Adds Something New

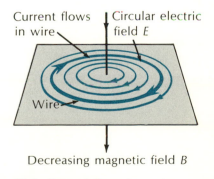

Current flows in wire

Circular electric field E

Wire

Decreasing magnetic field B

12-4 Current flows in a wire when the magnetic field changes.

So, without the help of any direct experimental verification, Maxwell proposed that the fourth equation be corrected to read:

> An electric field E which changes with time *as well* as a steady current I will set up a circular magnetic field.

With this new postulate, an electromagnetic wave theory was complete; Maxwell had proved that electromagnetic waves would have all the known properties of light waves. Using just Maxwell's equations, we can obtain all the phenomena of reflection, refraction, dispersion, and polarization for light. From them, Faraday's rotation of the plane of polarization by a magnet can also be predicted.

These equations represent an enormously successful theory and a beautiful synthesis of light with electricity and magnetism. However, the scope and applications of the theory were not appreciated until 1888 when Heinrich Hertz, who was attempting to verify Maxwell's theory, succeeded in producing and detecting electromagnetic radiation which exhibited a different frequency from light. His results led very quickly to the production and use of electromagnetic waves for communication: radio, shortwave radio, microwaves, radar, television, and the entire electronics communication technology. Figure 12-5 is a chart of the electromagnetic wave spectrum. Note that visible light occupies only a very small part of the spectrum (less than one octave, to use a musical analogy).

12-6

Maxwell's Equations without Magnetic Fields

We return finally to our original description of Maxwell's equations that eliminates the use of the magnetic field. The three equations below give a complete description of electromagnetism and are entirely equivalent to the usual five Maxwell equations where the magnetic field construct is used:

1 **Coulomb's Law:** $\qquad F = \dfrac{Q_1 Q_2}{r^2}\qquad$ for electrostatics

2 **Ampère's Law:** $\qquad F = \dfrac{e_1 e_2}{c^2 r^2}\qquad$ for electrodynamics

where $e_1 = I_1 \ell_1$ for a current or $e_1 = Qv$ for a charge moving with velocity v.

3 **Induction and radiation:** When a current element $e_1 = I_1 \ell_1$ is changed with time, either by changing the current I_1 or the geometrical configuration (represented by ℓ_1), an electric field is propagated from the current element into all space with the velocity of light. If the current element changes

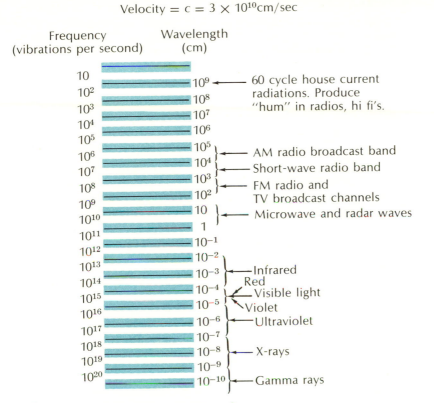

Velocity = $c = 3 \times 10^{10}$ cm/sec

Frequency (vibrations per second)	Wavelength (cm)	
10	10^9	60 cycle house current radiations. Produce "hum" in radios, hi fi's.
10^2	10^8	
10^3	10^7	
10^4	10^6	
10^5		
10^6	10^5	AM radio broadcast band
10^7	10^4	Short-wave radio band
10^8	10^3	FM radio and
10^9	10^2	TV broadcast channels
10^{10}	10	Microwave and radar waves
10^{11}	1	
10^{12}	10^{-1}	
10^{13}	10^{-2}	Infrared
10^{14}	10^{-3}	Red
10^{15}	10^{-4}	Visible light
10^{16}	10^{-5}	Violet
10^{17}	10^{-6}	Ultraviolet
10^{18}	10^{-7}	
10^{19}	10^{-8}	X-rays
10^{20}	10^{-9}	
	10^{-10}	Gamma rays

12-5 The electromagnetic spectrum of waves.

periodically with time, a periodically-changing electric field propagates at the velocity of light and obeys a wave equation: Periodic change in time of E = Periodic change in distance of E.

Note that the first of the three equations above is simply a restatement of Coulomb's law and the second is a restatement of Ampère's law (except that we have explicitly put in the velocity of light, the ratio of the electrostatic system of units to the electromagnetic system). Thus the current elements in the second equation should be in electrostatic units: I is in stat-A and Q is in stat-C, and the electromagnetic units (ab-C and ab-A) can now be discarded. The third equation describes induction and wave propagation: If there is an electrical conductor (an antenna) in space, the induced electric field or electric wave at the conductor (due to any of a myriad of electromagnetic wave sources) will cause a current in the conductor and thereby allow the reception of electrical pulses or patterns which can then be deciphered and used for communication.

12-7

The Implications of Maxwell's Revolution

Today the technological applications of Maxwell's revolution are evident everywhere around us. Our entire cultural and social structure is changing due to instantaneous (that is, at the speed of light) communications using electromagnetic radiation. However, the philosophical implications of this revolution did not really come to light until the late 1800's and are still not completely understood or appreciated even now. But the philosophical changes, like the technological changes produced by this revolution, are as striking as those of the Mechanical Revolution were 200 years before when the mechanistic picture of the universe replaced the organismic view of Aristotelian physics. This time, however, the mechanistic Newtonian universe was itself overthrown because of an entirely new outlook on nature, due partly to the continuing successes of the Maxwell theory and partly, as we will see in the next chapter, to the astounding results of one of the most significant single experiments in all of physics—the Michelson-Morley experiment.

Questions

1   What evidence led Maxwell to hypothesize that light and electromagnetic waves were one and the same phenomenon?

2   Do you share Maxwell's aversion to Newton's concept of "action-at-a-distance" forces in his gravitation theory? Explain.

3   Do you think the argument that nature should be symmetrical is sufficient reason to investigate the possibility of changing electric fields giving rise to magnetic fields?

4   Name some of the everyday objects or phenomena which operate based on Maxwell's electromagnetic waves.

5   Electromagnetic waves can be used to produce phenomena which involve energy. Do you think that electromagnetic waves violate the principle of conservation of energy or do they somehow involve energy?

6   If you decided in Question 5 that energy is somehow involved in electromagnetic waves, then speculate on the *momentum* content of electromagnetic waves. Is momentum in some way involved in these waves? Explain your answer.

7   Can you design a simple experiment to determine whether or not light has momentum?

8   Do you think that the changes in man's picture of the universe from the Electromagnetic Revolution were as sweeping as those which sprang from the work of Copernicus, Galileo, and Newton? Explain your answer.

9   Maxwell's predictions that electromagnetic waves existed and that light was composed of such waves led to a new question: In what medium do these waves propagate? All other waves propagate as disturbances of one kind or another in some kind of medium. Do

you think it inconceivable that light might propagate without a medium or do you think that a medium is necessary? Explain your answer.

10 If you feel that light waves need some medium in which to propagate, do you think that this medium would be detectable? Would the motion of the earth through this medium have any detectable effect? Explain.

Questions Requiring Outside Reading

1 Maxwell's achievements in the field of electricity and magnetism tend to overshadow his work in other fields. What were some of his contributions to the other branches of physics?

2 What similarities exist between the Electromagnetic Revolution and the Mechanical Revolution? How do the ways in which these two revolutions began differ?

3 Name some of the technological changes that have resulted from Maxwell's work.

4 How have these changes affected our culture? Our social and political scenes?

13

The End of the
Electromagnetic Revolution

In the last two decades of the nineteenth century, many famous physicists of the time were making very optimistic, if not completely ludicrous, statements on the state of physics. The general feeling was that everything, every phenomenon in the universe, was now completely known or at least knowable and predictable in principle. Newton's laws and Maxwell's equations gave man what he had been searching for over the centuries: the answers to all physical phenomena. The old and difficult problem of light had been shown to be a mere electromagnetic wave propagating on the ether. Old problems concerning the phenomena of heat and thermodynamics were also tamed; the kinetic theory of gases showed that thermodynamics and heat could be understood by applying Newton's laws to the statistical description of a very large number of particles—the molecules of a gas. The concepts of energy and energy conservation developed in the 1800's also permitted very elegant and mathematically high-powered reformulations of Newtonian mechanics, of which the theoretical physicists were understandably proud.

If everything discoverable had been discovered and everything knowable was known, the outlook for the physicists of the future looked rather bleak. Their only job would be to remeasure already known phenomena with finer and finer precision. In 1900, Lord Kelvin, one of the major contributors to the enormously successful theory of heat, collectively expressed the pride that all physicists felt over their newly acquired understanding of the phenomena of heat, light, and electromagnetism when he said that there were only two clouds obscuring "the beauty and clearness of the dynamical theory which asserts light and heat to be modes of motion."

One of the two clouds (in actuality, there were quite a few more) to which Kelvin alluded was the astounding result of the Michelson-Morley experiment in 1887. This provided not only unexpected philosophical implications of Maxwell's Electromagnetic Revolution, but, in addition, the experimental basis from which Albert Einstein was eventually to develop the theory of relativity.

Albert A. Michelson, an American physicist teaching at the U.S. Naval Academy at Annapolis, was refining experimental techniques begun by Thomas Young a century earlier to measure the interference from two rays of light. Michelson's experiment, completed in 1887 in collaboration with another American, Edward M. Morley, had a disarmingly simple purpose: To prove that, since light propagates on the ether with a velocity $c = 3 \times 10^{10}$ cm/sec, the velocity of the earth through this ether could be found by observing the velocity that light has on the earth as the earth is moving.

In Figure 13-1 the parallel lines represent the stationary ether on which a light disturbance propagates at a velocity $c = 3 \times 10^{10}$ cm/sec. The colored area is a table on the earth supporting a light source; the table moves through the ether with a velocity $\vec{v}$ to the right. An observer on the table at point (2) in the figure is separated from the light source by a distance d. If the earth were not moving through the ether [that is, if the source of light and the observer at (2) were both at rest in the ether], the time it would take the light to go from its source (2) would be equal to the distance d divided by the velocity c or $t = d/c$. However, since the earth moves with a velocity v, as the light travels on the ether from its source to (2), the observer at (2) is moving to the right *away*

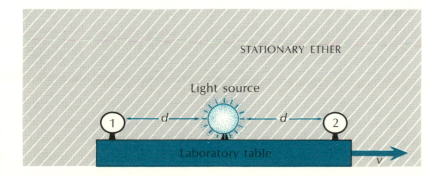

STATIONARY ETHER

Light source

1 ←— d —→ ←— d —→ 2

Laboratory table

v

13-1 The laboratory table (colored area) is moving to the right with a velocity $\vec{v}$ through the stationary ether. A light source on the table is received at both (1) and (2).

from the light waves as they approach him. Thus it takes longer for the light to reach (2) than it would if (2) were not moving through the ether. The velocity of light measured by the observer at (2) would, therefore, be smaller than c by the velocity v; that is, it would be measured as $(c - v)$. A similar analysis of the observer at (1) shows that his measurement of the velocity of light is $(c + v)$; the light reaches him sooner than it would if he were stationary, since observer (1) is moving with a velocity v *toward* the approaching light waves.

Unfortunately, because it is so large, direct measurements of the velocity of light c on our earth are not accurate enough to detect the small changes of $(c + v)$ or $(c - v)$ even if we assume some reasonable value for v. And what value of v would be reasonable anyway? If the earth were stationary in the ether (a most unlikely Aristotelian supposition), the daily rotation of the earth on its axis would yield a peripheral velocity v at the equator of about 4×10^4 cm/sec. This is only about $\frac{1}{1,000,000}$ of the velocity of light in the ether and is much too small to detect. If the sun were stationary in the ether, then the yearly motion of the earth around the sun would produce a tangential velocity v of about 3×10^6 cm/sec, which is approximately $\frac{1}{10,000}$ of the velocity of light—a fraction too small to detect in the 1800's.

<table>
<tr><td>

13-3

The Michelson-Morley Experiment

</td><td>

Michelson made clever use of his optical device, which he called an **interferometer**. He did not measure the velocity of light directly, but instead measured the *difference* between the velocity for light moving in one direction and the velocity for light moving in a perpendicular direction.

When the earth moves through the ether with velocity v, an observer (the experimenter) on the earth, considering himself to be stationary, must "feel" or experience an ether "wind" of velocity v coming at him from the direction toward which the earth is moving. In Figure 13-2, we assume the "wind" is moving by from left to right, as represented by the arrows. Light from the

</td></tr>
</table>

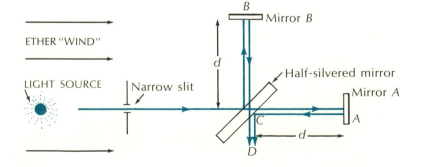

13-2 Interference of light in the Michelson-Morley experiment.

MICHELSON

Although born in Germany in December, 1852, Albert Michelson came to the United States with his family when he was two years old. The Michelsons settled first in the booming mining town of Virginia City, Nevada, moving to San Francisco a few years later. That city was then experiencing the tremendous rush which followed the discovery of gold in California in 1848. Michelson acquired his education in the public schools of the city while vigilantes were roaming the countryside, dealing out instant justice to all suspected of wrongdoing. □ He received an appointment from President Ulysses S. Grant to the Naval Academy and, after graduating, returned there to become an instructor in physics in 1875. His interest in physics grew as much at the Academy as his interest in regular Naval duties declined. By 1880, Michelson wanted to continue his study of physics and obtained a leave of absence to work in Germany and France. At 30, he finally resigned from the Navy and settled in Cleveland, Ohio, where he became a professor of physics at the Case School of Applied Science (now Case-Western Reserve University). □ After achieving much fame as a result of the world-renowned Michelson-Morley experiment, Michelson accepted the position of first Head of the Physics Department at the University of Chicago in 1892. Although interested in many fields of physics, he excelled in optics and interferometry. His interferometers were painstakingly made and revealed an ingenuity possessed by few other physicists. In 1907, he was awarded the Nobel Prize for his experiment with Morley as well as for his pioneering work in interferometry. □ During World War I, Michelson rejoined the Navy, but returned to Chicago when it ended. In his later years, he became interested in astronomy and used his light interferometers to measure the sizes of stars. This work so intrigued him that, at the age of 77, he resigned his post at Chicago to move to the Mount Wilson Observatory in Pasadena. Michelson died there in 1931 at the age of 79.

Albert Michelson

The Niels Bohr Library, American Institute of Physics

source passes through a narrow slit and is divided by a half-silvered mirror at point C in the figure. One light ray travels along path A to mirror A and is reflected back; the other light ray, traveling along path B, is reflected back by mirror B. The two rays then recombine at the half-silvered mirror C and the interference between them is observed at D in much the same way as the interference from two slits is observed in the double slit demonstration. The light along path A may be in phase, out of phase, or in some intermediate phase with the light from B, depending on the time it takes for light to travel from C to A versus the time it takes to travel from C to B.

Assume that ether wind is zero (that the experiment is not moving in the ether) and that paths CA and CB have the same length d. The time that it takes for light to go from C to mirror A is $t = d/c$; exactly the same amount of time is needed for the light to reflect back to C again. Thus the total time required for light to traverse path CAC is $t = 2d/c$. Likewise, the total time required for light to traverse path CBC is $t = 2d/c$. The times are the same, the two interfering rays are in phase, and a bright spot is found at the center of the interference pattern at D.

Now let the earth travel through the ether with a velocity v, so that the light in the experiment "feels" an ether wind (Figure 13-2). The light traveling from C to mirror A is moving in the ether with velocity c, but the ether itself is moving toward mirror A with velocity v. Thus the net velocity of light toward mirror A is $(c + v)$ and the time required to travel from C to A is $d/(c + v)$. In a similar manner, on returning from mirror A to C, the light travels against the ether wind, so the combined velocity is $(c - v)$. The time for the return trip AC is then $d/(c - v)$. The total time the light takes to travel along the path CAC is the sum of the times required to travel from C to A and from A to C: $d/(c + v) + d/(c - v)$ or, using $c^2 - v^2$ as the common denominator,

$$T_{CAC} = \frac{2dc}{c^2 - v^2} \qquad \text{(13-1)}$$

Now we want to compare this with the time t_{CBC} the light takes to travel to mirror B and back along a path at right angles to the ether wind. We want to find the combined velocity—the sum of the ether wind velocity and the velocity of light in the ether. To do this, we must use vectors since these two velocities are not in the same direction. The combined velocity $\vec{V}$ must be the *vector sum* of the velocity of light in the stationary ether $\vec{c}$ and the velocity of the ether wind $\vec{v}$. Furthermore, $\vec{V}$ must end up in a vertical direction (see Figure 13-3) so that the light will actually reach mirror B and not wind up on one side of it or the other.

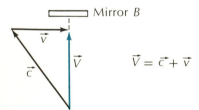

13-3 The velocity $\vec{V}$ of light moving perpendicular to the ether.

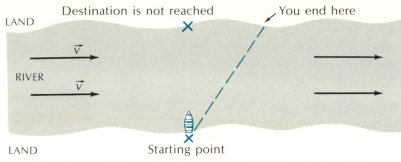

Destination is not reached   You end here

LAND

$\vec{v}$

RIVER

$\vec{v}$

LAND   Starting point

13-4 The boat will be swept downstream if it is headed straight across the river.

To see the problem in an everyday setting, imagine you are taking a motorboat across a river to a point on the far shore directly opposite you. The velocity of the river is $\vec{v}$ and the normal velocity of the boat in calm, stationary water is $\vec{c}$. If you aim the boat directly toward the opposite point on the far shore, of course it will be swept along to the right (see Figure 13-4) by the water current as it travels across and will end up downstream from its appointed destination. In order to reach a directly opposite point on the far shore, the boat must be aimed *upstream* as shown in Figure 13-5 in a way that compensates for the downstream motion of the water.

The actual, effective velocity (relative to the land) of the boat from its starting point to its destination point is less than the velocity c of the boat in calm water. This actual, effective velocity can be found by using the Pythagorean Theorem for the right triangle formed by the velocities (shown in Figure 13-5):

$$c^2 = v^2 + V^2$$

or

$$V = \sqrt{c^2 - v^2}$$

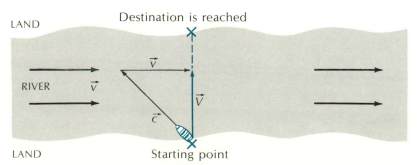

Destination is reached

LAND

$\vec{v}$

RIVER $\vec{v}$ $\vec{V}$

$\vec{c}$

LAND   Starting point

13-5 To reach a directly opposite point on the far shore, the boat must be headed a sufficient distance *up*stream to compensate for the river's downstream motion.

Using the analogy of the boat and the river, the velocity of light in the Michelson-Morley experiment along the path perpendicular to the ether wind CB is $V = \sqrt{c^2 - v^2}$, so that the time to traverse the distance d from C to B in Figure 13-2 is

$$t_{CB} = \frac{d}{\sqrt{c^2 - v^2}}$$

The time for the return trip from B to C must be the same, so that the total time that light takes to travel along path CBC is

$$t_{CBC} = \frac{2d}{\sqrt{c^2 - v^2}} \qquad \text{(13-2)}$$

which does not look the same as that for path CAC given in Equation (13-1). The interference of the two rays depends on their time difference and, when we express t_{CBC} in another form directly comparable to t_{CAC}, we find that

$$t_{CBC} = t_{CAC} \sqrt{1 - \frac{v^2}{c^2}} \qquad \text{(13-3)}$$

The time light takes to travel perpendicular to the ether (t_{CBC}) is *shorter* by a factor of $\sqrt{1 - v^2/c^2}$ than the time light takes to travel parallel to the ether (t_{CAC}). Now if we assume this time difference corresponds to one-half of a vibration of the waves, for example, the light arriving at the center of the interference pattern from path A will be lagging behind the light from path B by one-half of a vibration or one-half wavelength. Thus the two rays will be out of phase and a dark spot will be seen at the center of the interference pattern at D.

In the actual experiment Michelson performed, the entire interferometer apparatus was rotated 90° (see Figure 13-6) so that the path CAC became perpendicular to the ether wind and path CBC became parallel to it. With the interferometer in this new position, the time the light takes to traverse the path CAC is shorter than the time it takes to traverse the path CBC. Assume that, before rotating the apparatus, the light from path A was behind the light from path B by one-half vibration and that somewhere halfway through the rotation, the light from A is exactly in phase with the light from B. On viewing the center of the interference fringes, we *should* first see a dark spot out of phase, then a light spot in phase, and finally another dark spot as the apparatus is rotated. Knowing the time difference between the two paths and the frequency of the light used, the velocity v of the earth through the ether can be easily calculated using Equation (13-3).

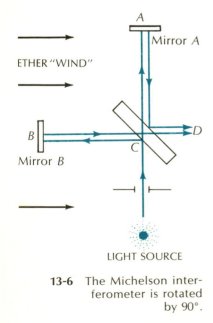

13-6 The Michelson interferometer is rotated by 90°.

The unexpected, startling discovery Michelson and Morley made was that, as they rotated the apparatus, there was *no change in the interference pattern*. They repeated the experiment innumerable times, as did many other physicists, always with the same negative results. If there was no change in the interference pattern, then the only conclusion could be that the ether wind velocity (our velocity through the ether) is *zero*. The velocity of the earth through the ether is evidently zero under all experimental conditions—it does not depend in any way upon the velocity of the earth about its own axis or the velocity of the earth around the sun.

The Aristotelian-Ptolemaic universe with a motionless earth in the ether would have accounted for these negative results, but, nineteenth century physicists could no longer seriously consider this explanation. The only other possible explanation they could provide for such a result was that perhaps a small amount of ether was carried along with the earth in its travels through space. This theory, called the ether drag theory, was tested in many ways. Light waves from distant stars propagating toward the earth on the stationary ether should be bent as they enter the moving ether attached to the moving earth. The results of experiments on the aberration or the bending of starlight showed that there could be no ether drag.

After attempting to avoid the inevitable, physicists finally came face to face with the inescapable conclusion that if there were no possible way to detect our velocity through the ether, then *there was no ether* and the equations for t_{CAC} and t_{CBC} were representative of situations which *did not actually exist*. The ether was originally invented as a construct to provide a medium or a mechanism for the propagation of light waves as an analogy to water for water waves, air molecules for sound waves, and so on. However, the Michelson-Morley experiment showed that the ether construct was not simply unnecessary, but that its presence would predict a result contrary to their experimental findings.

13-4

The Results of the Michelson-Morley Experiment

13-5

The Philosophical Consequences of the Electromagnetic Revolution

If there is no ether, no medium on which the electric and magnetic disturbances of light propagate from one place to another, how do we "explain" light? This question, when scrutinized closely, focuses on the central point in the philosophical revolution that followed the development of Maxwell's equations—a revolution of the greatest importance to the philosophy of science. Although just as startling and influential as the Newtonian Mechanical Revolution, the "Maxwellian Revolution" has not been adequately understood and interpreted by many writers in the history and philosophy of science.

The fact is that Maxwell's equations work; they describe nature extraordinarily well, in spite of the "disappearance" of the ether. We conclude that the equations of electromagnetism stand on their own empirical merit, regardless of the lack of the mechanistic basis of the ether. With the negative result of the Michelson-Morley experiment alone, the edifice of the Newtonian mechanical universe began to crumble. Aristotle had stressed the importance of man by stating that the earth was at the center of the universe; the work of Copernicus, Kepler, and Galileo to change this view culminated in Newton's mechanistic concept of the universe. In the 1800's the reintroduction of the concept of the ether provided a "center" of the universe (that is, a state of motion for which the velocity of light was exactly c, where Maxwell's equations held). In other **reference frames** moving with respect to the ether, it would have been necessary to change Maxwell's equations because the measured velocity of light would have been different from c. But Michelson and Morley showed that the velocity of light was not in fact different—that it was the same no matter what velocity the earth-laboratory had in space.

The Michelson-Morley experiment revealed that there was no privileged position or "center" of the universe even in the sense of privileged motion, since there is no ether. But, more importantly, the experiment invalidated the mechanistic props on which Faraday's and Maxwell's electromagnetic theory was predicated. Since the Maxwell theory still works, with or without the ether, we can only conclude that *any mechanistic explanation of Maxwell's theory has to be discarded.* The equations still describe nature, even if we cannot visualize or conceive the propagation of electromagnetic waves in the absence of a medium that can be stressed, compressed, or stretched to transmit the disturbances.

Although Maxwell personally believed in and tried to work out a mechanical explanation of his equations, he was never successful. He did believe that a mechanistic explanation or picture would be developed eventually and, in actual fact, a large number of theoretical physicists did occupy themselves with this problem in the latter part of the nineteenth century. Maxwell, like Copernicus, Galileo, and Kepler (described by Arthur Koestler as "sleepwalkers"[1] in his book of the same name), did not realize the enormous importance his own discovery would have on the future, both in the technology it produced and also in the contribution it made toward changing the Newtonian world-picture.

The fact that we cannot relate Maxwell's mathematical laws to our previous mechanistic sense experiences is unnerving to

[1] Arthur Koestler, *Sleepwalkers* (New York: Macmillan, 1959).

some people and even demoralizing to others, but perhaps it should not be surprising. What authority demands that everything in nature must be pictured or understood in a mechanistic way through our accumulated sense experiences? This new principle will become more and more important as we learn about the Particle Revolution.

Questions

1 Do you find the idea that light travels as a wave without a medium strange? Explain.

2 Describe the Michelson-Morley experiment and explain verbally how it should have detected the ether "wind."

3 Can experiment prove a theory? Explain.

4 Can experiment disprove a theory? Explain.

5 With respect to what frame of reference was the speed of light thought to be constant prior to the Michelson-Morley experiment? Why were the results of the Michelson-Morley experiment so incomprehensible to the scientists of the time?

6 For what reasons did the Michelson-Morley experiment confound physicists in the late nineteenth century? Was the fact that it failed to detect the ether necessarily a cause for consternation? Or was it possible for scientists to ignore this one shortcoming since Maxwell's equations worked so beautifully in spite of it?

Questions Requiring Outside Reading

1 In what ways can Copernicus, Galileo, and Kepler all be considered "sleepwalkers?"

2 Read about some of the attempts to explain the failure of Michelson and Morley's experiment. In what ways did these explanations fail? (A good readable source for such material is *The Theory of Relativity and Principles of Modern Physics* by Huseyin Yilmay, Blaisdell [1965]. The treatment of the Michelson-Morley experiment is elementary and nontechnical.)

3 Why did scientists invest the concept of the ether? (i.e., for what reasons was the ether believed to be necessary?)

4 Describe some of the supposed properties of the ether. Are any of these properties inconsistent with one another?

5 When the Michelson-Morley experiment failed to detect the motion of the earth through the ether, theoretical physicists were confronted with several alternatives to explain this failure. What were some of these alternatives and what were the objections to them? (*Hint:* Read the next chapter.)

14

The Beginning of the
Particle Revolution:
Einstein and Relativity

14-1

Introduction

The Michelson-Morley experiment provided a new philosophical way of looking at a universe that could no longer be explained by the old mechanistic pictures. It also initiated a new revolution in physics. This third revolution, which we will call the Particle Revolution, is still incomplete and its resolution, the next grand synthesis, remains in the future. Three general fields of physics presently comprise this "modern" revolution: special relativity, quantum mechanics, and, the most current field of research, elementary-particle physics (or high-energy physics) for which the physical "laws" are still undiscovered. Most university research in physics today is either in particle physics or in applications of quantum mechanics. These latter applications are part of a "mop-up" period in quantum mechanics and consequently are quite diverse; they range from the study of atomic nuclei (nuclear physics) to the study of atoms (atomic physics) to molecules (quantum chemistry) to the properties of gross matter (solid state physics and low temperature physics).

14-2

The Bases for Einstein's Relativity

Historically, Einstein's relativity work appears to be a natural consequence of the null result of the Michelson-Morley experiment: that the velocity of light was always the same no matter what velocity the observer (the laboratory) had in space. It may be concluded from this that the measured velocity of light will be the same to *all* observers no matter what their relative velocities may be. Einstein put this in the form of his first basic postulate:

The velocity of light is independent of the motion both of the observer and of the source (for inertial reference frames).

This seems to be a completely understandable, common-sense statement. But, as we will see, it leads directly to some very uncommon results.

A second basis of relativity was probably even more important to Einstein than the result of the Michelson-Morley experiment. Maxwell's first two laws are, respectively, Coulomb's law for stationary, nonmoving charges and Ampère's law for moving charges (current-current force). Coulomb's law

$$F_{Coul} = \frac{Q_1 Q_2}{r^2}$$

holds for two charges Q_1 and Q_2 at rest; if the two charges are both moving with a velocity v, then each charge constitutes a current element Qv and the force due to Ampère's law

$$F_{Amp} = -\frac{1}{c^2} \frac{(Q_1 v)(Q_2 v)}{r^2}$$

must be added to the force from Coulomb's law. This means that if a man moving at a constant velocity v on a railroad train, for example, measures the force between two charges Q_1 and Q_2 that are also on the train, he will measure only the force due to Coulomb's law F_{Coul}. However, another man on the ground who sees the train and the charges Q_1 and Q_2 moving past him with a velocity v will find the force due to Coulomb's law and, in addition, a force due to Ampère's law for moving charges and measure the force between the two charges as $F_{Coul} + F_{Amp}$.

Why don't the moving observer and the stationary observer measure the same total force? Copernicus, Galileo, and Newton each postulated the **principle of relativity**—that the laws of physics would be the same to an observer moving with a constant velocity v as they would be to a stationary observer. In other words, *there is no physical experiment to detect one's state of motion at constant velocity.* Therefore, there is no "absolute" motion or lack of motion; *all motion at constant velocity is relative.*

Newton's laws manifestly agree with this principle of relativity. His force equation for gravitation

$$F = G \frac{m_1 m_2}{r^2}$$

does not contain any velocity. Thus the gravitational force is the same to all observers, independent of their state of motion at

EINSTEIN

Albert Einstein was born in 1879 in the small southern German town of Ulm. No child prodigy, he was considered a poor student and irritated his teachers with his deliberate slowness—due probably to both his dislike of the school routine and his balking at the tedium of rote learning. By the time he was 11, however, Einstein was reading mathematics and philosophy books. He also began learning the violin as a boy, keeping up the music for the rest of his life. □ At the age of 17, Einstein entered the Polytechnic Institute in Zurich to be trained as a teacher in mathematics and physics. He was helped financially by some rich relatives and tutored students as well. He received his diploma in 1901 and, unable to find a suitable academic or teaching position, took a job inspecting patents for the Swiss government in Berne. During his time in Berne, he worked privately on physics, married, and had two sons. □ The year 1905 marked Einstein's emergence as a physicist, and his work at that time—he wrote four fundamental papers and also received his doctoral degree—destined him to be a giant in the field. His work on relativity, although slow to be accepted by established physicists of the time, was eventually to make him famous. He received a professorship at Zurich in 1909, the Chair of Theoretical Physics at Prague in 1911, and, finally, the Chair at the University of Berlin in 1914. □ Einstein's personal life suffered as much as his professional life prospered. In the move to Berlin, he separated from his wife and the marriage was later dissolved. He was deeply upset about leaving his sons; according to one biographer, it was probably the only time that anyone ever saw Einstein cry. □ While living in Berlin, Einstein worked on the General Theory of Relativity; his predictions on the bending of starlight by the sun were verified by English astronomers in 1919. The President of the Royal Society called Einstein's theory one of the "greatest achievements in human thought." □ Einstein's cousin, Elsa, was also living in Berlin and the two became very close. He took her as his second wife in 1917. Elsa was a strong, compassionate woman who helped make her husband's everyday life easier and allowed him to retreat safely and comfortably into his private world of physics. □ Einstein was awarded many honors in his life, notably the Nobel Prize in 1921 for his work on the photoelectric effect. In his later years, he worked unsuccessfully on a unified field theory that attempted to bring together all of electromagnetism with gravitation. He also spent time writing and lecturing both on a world government and on Zionism. At the time of Hitler's rise to power in Germany, Einstein—a Jew—left for America, writing "As long as I have any choice, I will only stay in a country where political liberty, tolerance, and equality of all citizens before the law prevail.—These conditions do not obtain in Germany at the present time."[1] □ Einstein lived out his later years in Princeton, New Jersey, and died there on April 18, 1955. He was 76.

[1] Albert Einstein, *Ideas and Opinions* (New York: Crown Publishers, Inc., 1954), p. 205. Used by permission of Crown Publishers, Inc.

constant velocity. Furthermore, Newton's second law $\vec{F} = m\vec{a}$ tells us that forces produce accelerations or *changes in velocity*, and it is easy to see that a *change in velocity* will be measured identically by all observers no matter what their velocities relative to each other were before the force was applied. However, Maxwell's equations, as we have seen, seemed to depend upon the observer's velocity of motion and therefore to violate such ideas about the relativity of motion. This was the problem that Einstein tackled

Albert Einstein in his study.

Ernst Haas, Magnum

in 1905 with his now legendary paper proposing what he called the Special Theory of Relativity.

The dilemma that Einstein was trying to solve was the contradiction between the descriptions of nature implied in Maxwell's equations and in Newton's laws. Maxwell's equations clearly violated the content of what we will call *Galilean relativity*, since the electrodynamic forces we actually measure do indeed depend on our state of motion. There would seem to be three alternatives to reconcile this dilemma:

1 Throw out Maxwell's equations and attempt to find a new electrodynamic theory in which velocities (relative motion) do not explicitly appear.
2 Abandon the entire *Principle of Relativity* and assume that the laws of physics are different when viewed by observers in different states of relative motion (constant velocity). Thus, there would be a preferred state of motion in the universe.
3 Retain the *Principle of Relativity*, but throw away the specific mathematical form of this principle—Galilean relativity—and replace it with a new form.

Maxwell's equations were so successful that no one was ever able to accomplish the first alternative; many new theories were tried, but none came as close to describing nature and to fitting the electromagnetic data. The third alternative was unthinkable; discarding Galilean relativity would mean, in effect, throwing out all Newtonian mechanics and replacing it with something new. So, until the beginning of the twentieth century, the second alternative was accepted; a justification for this was the assumed existence of the ether—the preferred state of the universe in which Maxwell's equations held.

When the ether construct was destroyed by the Michelson-Morley experiment, the problem arose again. Einstein turned to the third alternative, probed deeper than any before him, and succeeded in overthrowing all our "common-sense" ideas on space, time, and motion. He took the Principle of Relativity:

All inertial reference frames are equivalent: The laws of physics have the same form.

as his second basic postulate. In his 1905 paper, Einstein stated:

> The unsuccessful attempts to discover any motion of the earth relative to the "light medium," suggest that the phenomena of electrodynamics as well as of mechanics possess no properties corresponding to the idea of absolute rest. They suggest rather that, as has already been shown to the first order of small quantities, the same laws of electrodynamics and optics will be valid for all frames of reference for which the equations of mechanics hold good. We will raise this conjecture (the purport of which will hereafter be called the "Principle of Relativity") to the status of a postulate, and also introduce another postulate, which is only apparently irreconcilable with the former, namely, that light is always propagated in empty space with a definite velocity c which is independent of the state of motion of the emitting body. These two postulates suffice for the attainment of a simple and consistent theory of the electrodynamics of moving bodies based on Maxwell's theory for stationary bodies. The introduction of a "luminiferous ether" will prove to be superfluous inasmuch as the view here to be developed will not require an "absolutely stationary space" provided with special properties.[2]

14-3

Galilean Relativity

Galilean relativity expresses the Principle of Relativity in a common-sense, mathematical form: The laws of physics are the same to a stationary observer A as they are to an observer A' who is moving with a constant velocity $\vec{v}$ relative to observer A (as in

[2] Albert Einstein, "On the Electrodynamics of Moving Bodies," *Annalen Physik* 17 (Leipzig, 1905): 891. English translation from Albert Einstein *et al.*, *The Principle of Relativity* (New York: Dover Publications, Inc., 1924), pp 37–38.

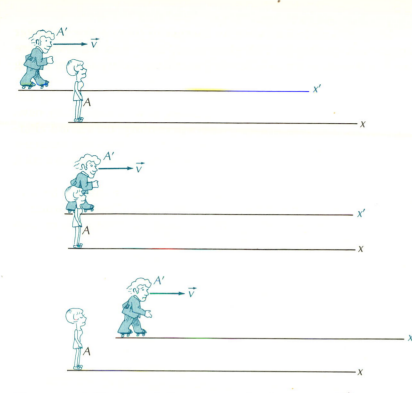

Figure 14-1). Horizontal distances in the direction of $\vec{v}$ are measured by A along the direction x and by A' along the direction x' in the "moving" reference system. Consider x as a very long ruler fixed to A and x' as another very long, *moving* ruler fixed to A'. At a later time, A' is at the same place as A and, at a still later time, A' moves to the right of A. The mathematical form of Galilean relativity is merely the relation of distance x' measured by A' in the moving system to the distance x measured by A in the stationary system. If the stationary observer A measures his distance from some object as x_1, then A' in the moving system will measure a distance x_1' from the same object. How is x_1 related to x_1'? From Figure 14-2, it is clear that

$$x_1 = x_1' + vt \qquad (14\text{-}1)$$

where vt is the distance between A' and A for a constant velocity

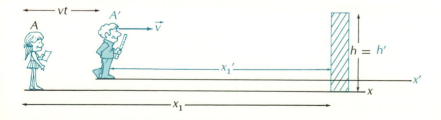

14-2 Observer A measures a distance x_1 from an object; A' measures x_1' from the same object.

v. Note that we measure the time t from the instant A' overtakes A. Equation (14-1) gives the transformation of distances measured by A' to the distances measured by A. This was called the **Galilean transformation**, since it expressed the common-sense ideas on relative motion originally set forth by Galileo and Newton.

Distances measured in other directions perpendicular to the direction of the velocity v obviously will be measured as the same by both A and A'. Thus, the height h' measured by observer A' in Figure 14-2 will be the same as the height h measured by observer A.

Galilean relativity also assumes implicitly that the time t when observer A measures his distance x_1 from the object is the same as the time t' when observer A' measures x'_1. If the times t and t' were not equal, Equation (14-1) would make no sense. We assume then that A and A' can make *simultaneous* measurements of their respective distances x_1 and x'_1 from the object. This is the second equation of Galilean relativity

$$t = t' \qquad\qquad\qquad\qquad \textbf{(14-2)}$$

These two equations are the mathematical statement of the Principle of Relativity in which physicists believed until 1905. It is now evident that Newton's laws for forces are measured the same for observer A and observer A'. Newton's gravitational force

$$F = G\,\frac{m_1 m_2}{r^2}$$

depends upon the square of the distance r between two masses. This separation r, dependent on the difference between two distances, would be measured the same by A as by A'; thus the force would be the same. Newton's second law $\vec{F} = m\vec{a}$ or $\vec{F} = m(\Delta\vec{v}/\Delta t)$ would be the same for A as for A' since the force $\vec{F}$ does not depend on the velocity of motion $\vec{v}$, but rather on the *change in velocity* $\Delta\vec{v}$. Therefore, as long as A' has a constant velocity and does not change his velocity (that is, does not accelerate) with respect to A, Newton's second law is the same for both A and A'.

However, Maxwell's equations, which explicitly contain the velocity v of a moving charge, present a problem. Using them, the force between two charges is measured differently by A than by A', seems to conflict with the Principle of Relativity. Yet, Maxwell's equations and, in particular, Ampère's law are experimentally observed; they are correct descriptions of nature. How then can we resolve this paradox? We must return to the third alternative presented and assume that the Principle of Relativity is still valid, but that the common-sense, mathematical statement of the principle—the Galilean transformation—is incorrect.

The clue that Einstein found for a remedy to this situation comes from a careful examination of Equation (14-2), $t = t'$, which states that the time t of some event measured by A is the same as the time t' of that event measured by A'. It also states that the time difference of two events measured by A is the same as the time difference of the same two events measured by A'; that is, two events which are measured as simultaneous by A will also be measured as simultaneous by A'. But this is clearly not true if the means by which the events are observed take a finite or nonzero amount of time to travel to each observer. To illustrate, let us take light, which has the fastest known velocity. Observer A is stationary on the ground and observer A' is moving past with a velocity v. At the instant A' passes A, two light flashes occur, one to the left at point B and one to the right at point C (see Figure 14-3 a). Observers A and A' are equidistant from the flashes at this time. As the light waves travel outward with a velocity c, observer A' continues to move to the right with velocity v. The two light waves will reach observer A at the same time and he will consider them to be simultaneous. However, A' is still moving toward point C and will observe the light flash from C *before* he observes the flash from B. Thus the two events, the two flashes of light, are not simultaneous to A' (Figure 14-3 b). The statement of Galilean

14-4

Simultaneity

Light flash
B

Light flash
C

— d —   — d —

14-3a Observers A and A' are equidistant from two light flashes which occur at B and at C.

Light wave from B

Light wave from C

B

C

14-3b A' will observe the light flash from C before he observes the flash from B.

relativity $t = t'$ can be exposed as manifestly false even by these common-sense arguments. In some way, the time t' measured by A' must depend not only upon the time t measured by A, but also upon the velocity of relative motion v.

Our conclusion from this example is best given in Einstein's own words:

> So we see that we cannot attach any absolute signification to the concept of simultaneity, but that two events which, viewed from a system of coordinates, are simultaneous, can no longer be looked upon as simultaneous events when envisaged from a system which is in motion relatively to that system.[3]

A new transformation must therefore replace the incorrect Galilean transformation equations (14-1) and (14-2). This can be obtained by returning to Einstein's basic postulate which he based on the null result of the Michelson-Morley experiment: *The velocity of light is always the same and does not depend on the motion* (at constant velocity) *of the observer*. First we must express this statement in mathematical form and then try to obtain from it a new set of transformation equations to replace Equations (14-1) and (14-2).

14-5

The Basic Equation

If observer A flashes a light, in time t the light will have traveled a distance $x = ct$, where c is the velocity of light. To account for the fact that x could be measured either as positive to the right or negative to the left (see Figure 14-4), we square each side of $x = ct$ and write

$$x^2 = c^2 t^2 \tag{14-3}$$

as the equation describing the distance x traveled by light in time t as *observed by* A.

Now let us consider how this same event is observed by A', who is moving past A with velocity v. Assume that at the instant A' passes A, the light flash is set off by either or both observers. The moving observer A' will measure a distance x' in his own coordinate system for the light to travel in time t' (the time as measured by A'). But, according to our interpretation of the Michelson-Morley experiment, the velocity of light is *independent of the motion of the observer*. Therefore, A' must also measure the light flash as traveling at a velocity c in his "moving" system since, according to him, he is stationary and A is moving by him

[3]Albert Einstein, *The Meaning of Relativity*, p. 30. Copyright 1953, Princeton University Press; 1956, estate of A. Einstein. Used by permission of Princeton University Press.

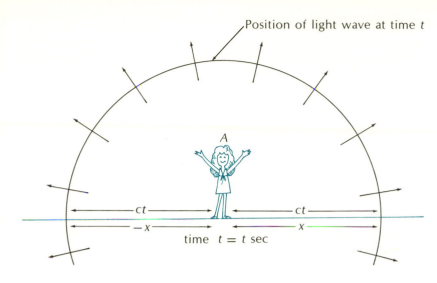

Position of light wave at time t

A

ct

ct

−x

x

time t = t sec

14-4 After t sec, the flash has traveled a distance $x = ct$ to the right and $-x = ct$ to the left.

with velocity v. A′ would therefore have to write an equation analogous to Equation (14-3),

$$x'^2 = c^2t'^2 \qquad \text{(14-4)}$$

where the distance x′ and time t′ are both measured by A′. The last step in our algebra is to move c^2t^2 from the right-hand side of Equation (14-3) to the left-hand side: $x^2 - c^2t^2 = 0$; we do the same for $c^2t'^2$ in Equation (14-4): $x'^2 - c^2t'^2 = 0$. Since the quantities $(x^2 - c^2t^2)$ and $(x'^2 - c^2t'^2)$ are each equal to zero, they must also be equal to one another. Thus, we arrive at the mathematical formulation of the statement that the velocity of light is independent of the observer's motion at constant velocity.

$$x^2 - c^2t^2 = x'^2 - c^2t'^2 \qquad \text{(14-5)}$$

This equation can be taken as the fundamental "law" of relativity. Any new transformation equations between x and x′ and t and t′ which replace the Galilean transformation equations must satisfy or agree with Equation (14-5).

It is important to test Galilean relativity or the Galilean transformation by applying Equation (14-1) to Equation (14-5). We substitute $x - vt$ for x′ in Equation (14-5) and express everything in terms of x and t. Then we substitute Equation (14-2), $t = t'$, to eliminate t′ in favor of t. When we work this new equation out, we find $x = vt/2$.

But this result makes no sense since it says that the distance x traveled by the light flash is one-half the distance vt observer A′ travels from A. Since the velocity v of A′ relative to A is

14-6

The Failure of Galilean Relativity

completely arbitrary and has no connection with the velocity of light, we conclude that the result $x = vt/2$ is nonsense or a contradiction and that the Galilean transformation equations we used for space (14-1) and time (14-2) are inconsistent with the basic postulate of relativity, Equation (14-5).

14-7

The Lorentz Transformations

The transformations of distance x to distance x' and of time t to time t' can be obtained directly from Equation (14-5) by means of a little algebra. Instead of deriving these transformations directly here (they can be found in many general physics textbooks), we will simply state the transformation equations and show in Appendix 14 (page 304) that they do indeed satisfy the fundamental law of Equation (14-5).

The transformation for distances that replaces Equation (14-1) is

$$x = \frac{x' + vt'}{\sqrt{1 - v^2/c^2}} \tag{14-6}$$

and the transformation for time, replacing Equation (14-2) is

$$t = \frac{t' + vx'/c^2}{\sqrt{1 - v^2/c^2}} \tag{14-7}$$

Compared with the earlier, incorrect Galilean transformation equations (14-1) and (14-2), equations (14-6) and (14-7), called the **Lorentz transformations**,[4] are certainly more complicated; however, they are not as formidable as they first appear to be.

The transformations for x' (in terms of x and t) and for t' (in terms of t and x) can also be written. Since we believe in the principle of relativity, the equations for x' and t' should look the same as Equations (14-6) and (14-7) for x and t if all primed symbols (denoting the reference system of A') were interchanged with unprimed symbols (denoting the reference system of A) and the sign of the relative velocity v was changed to $-v$. In other words, the transformations of distances and times from the unprimed, stationary system A to the primed moving system A' must be the same as the transformations from A' back to A, except that the relative velocity will be in the opposite direction. Therefore,

$$x' = \frac{x - vt}{\sqrt{1 - v^2/c^2}} \tag{14-8}$$

[4] Named after H. A. Lorentz, the famous Dutch physicist who worked on the theory that electricity is carried by submicroscopic particles (which we now call electrons). Lorentz's work on a relativity theory, although done independently, produced results that came close to Einstein's.

and

$$t' = \frac{t - vx/c^2}{\sqrt{1 - v^2/c^2}}$$

(14-9)

In Appendix 14 (page 304) we show that these relativistic transformations of space and time are indeed a solution of the basic law of relativity, Equation (14-5).

The experimental fact that forms the basis of relativity is that the velocity of light always has the same measured value, no matter what motion at constant velocity an observer or the light source may have. This may seem like a plausible statement at first sight, but, on reflection, it conflicts with our everyday, common-sense experiences. Normally, in the motion of material objects, we think of velocities as additive.

If a cyclist travels at 10 mph on the street and a person runs along behind him at 9 mph, then, relative to the runner, the cyclist is moving away at only 1 mph—the difference between the two velocities. As another example, if two cars, each traveling 60 mph, are approaching one another head-on, the speed of one car relative to the other will be the sum of their velocities or 120 mph. Our day-to-day life is full of examples like these.

But, according to the results of many, many experiments, light obeys different rules. A light beam passes by an observer at 186,000 miles per second and he tries to "catch up" with it by traveling in a rocket ship (relative to his original frame of reference) in the same direction at near the velocity of light—say, at 185,000 miles per second. Can he measure that the original light beam recedes from him at $186,000 - 185,000 = 1,000$ miles per second (the difference between the two velocities)? No, he must still measure the velocity as 186,000 miles per second relative to him, no matter what velocity *he* has.

All observers, no matter what their velocities, measure the same speed for light. Light from a flashlight aimed at you from the opposite end of a room will pass by you at 186,000 miles per second, as measured by you at rest in the room. If the flashlight then begins to move toward you ar 100,000 miles per second while still emitting light at 186,000 miles per second, you will not measure the light's speed as the combined speed of 286,000 miles per second, but simply as 186,000 miles per second. The person moving toward you with the flashlight will also measure the light leaving the flashlight at 186,000 miles per second. Again, the velocity of light is always measured as the same value by all observers (186,000 miles per second in air or in a vacuum) and is not affected by the observer's motion or the source's motion.

14-8

The Constancy of the Speed of Light

It may take a while to become accustomed to this novel property of light, since it has no basis in any of our previous sense experiences. Remember, too, that this property is not simply theory or conjecture; it has been borne out experimentally by every test physicists have proposed to measure the velocity of light. We will use it in the next chapter to derive some unexpected and unusual consequences of Einstein's special relativity.

Questions

1 In what sense does Einstein's Special Theory of Relativity render the ether concept "meaningless?" After Einstein, why was it no longer necessary to look for the motion of the earth through the ether? Explain.

2 Did the Special Theory of Relativity really prove that the ether was nonexistent or did it simply eliminate the need for postulating the ether? Explain.

3 State the Principle of Relativity according to Einstein. According to Galileo. What is the difference between the two?

4 Explain how the equations of the Lorentz transformations enable different observers to reconcile seemingly contradictory observations on the velocity of light.

5 What physical value does *not change* from one observer to another in special relativity?

6 The equations of special relativity imply that there is no such thing as absolute simultaneity. Do you think that it is still valid to use the concept of simultaneity in science? Explain your answer.

7 Do you think that Einstein's revision of man's picture of the universe was as sweeping as the revolution produced by Copernicus, Galileo, and Newton? Explain.

8 Do you think that there could be physical laws which the human mind could not, in principle, understand? Explain. (Do not limit yourself to ideas that are simply too complicated or are too great in scope to be grasped all at once.)

Questions Requiring Outside Reading

1 Discuss some of the reasons that the Special Theory of Relativity and the Lorentz transformations remained undiscovered for so long a time.

2 What is the nature of time in Einstein's picture of special relativity? How is time related to other quantities such as distance?

3 Why do we refer to time as the fourth dimension? Is time really a "dimension"—another direction along which to measure positions?

4 Discuss the contributions Poincaré, Lorentz, and Fitzgerald made to relativity theory before 1905.

1   Observer A sees an event take place at the position $x = 200$ cm, $t = 3 \times 10^{-8}$ sec. Observer B is moving with velocity $v = 0.6\ c$ along direction x. Write the *transformation equations* which will tell us how to find the position and time at which B observes this event. Then use these equations to find the values x' and t' which B will measure.

2   Write the Galilean transformation equations for the situation in Problem 1. What are x' and t' as calculated for Galilean relativity?

3   The square root $\sqrt{1 - v^2/c^2}$ appears frequently in the Lorentz transformations. What happens to the value of this number as c becomes infinite or as v becomes minute in comparison with c? What happens to the transformation equations in each of these cases?

4   What happens to the value of the square root in Problem 3 if v closely approaches c in value? If v equals c?

15

The Consequences of Relativity

As we have seen in the last several chapters, the fact that different observers moving at constant velocity with respect to one another each measure the *same* velocity for light led to drastic changes in our old, common-sense ideas on time, space, mass, and energy. To begin to see why this is true, we will consider two observers A and A', moving past each other with a relative velocity v (since motion is only relative, we cannot consider either observer to be "at rest"). A light flashes for an instant at the precise moment A and A' pass one another. A short time t later, observer A will measure the resulting sphere of light as centered around himself and moving outward with velocity c. However, A', by now a distance vt away from A, will also measure the same light sphere as centered around himself (A') moving outward with velocity c. How can two observers A and A', separated by a distance vt from one another, each measure that they are at the center of the *same* sphere of light? The answer is that measurement of an object's space (distances) and its time must change when the object is moving with respect to the observer. In Einstein's words:

> The theory to be developed is based—like all electrodynamics—on the kinematics of the rigid body, since the assertions of any such theory have to do with the relationships between rigid bodies (systems of coordinates), clocks, and electromagnetic processes. Insufficient consideration of this circumstance lies at the root of the difficulties which the electrodynamics of moving bodies at present encounters.[1]

[1]Albert Einstein, "On the Electrodynamics of Moving Bodies," *Annalen Physik* 17 (Leipzig, 1905): 891. English translation from Albert Einstein *et al.*, *The Principle of Relativity* (New York: Dover Publications, Inc., 1924), pp. 37–38.

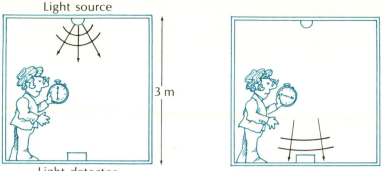

Light source

3 m

Light detector

15-1 A light on the ceiling starts the stopwatch; a detector on the floor stops the watch.

The first new result of Einstein's special relativity which we will prove is that we measure time (mechanical and electric clocks as well as atomic and biological processes) in a reference frame moving past us with velocity v as *slower* than time in our own reference frame.

Observer A is in a room 3 m high with a light source (a flash bulb) on the ceiling and a light detector (an electric eye) on the floor directly below the bulb. A flashes on the light, which starts a stopwatch that measures time in **nanoseconds** (10^{-9} sec; abbreviated nsec); when this same light flash hits the detector, the watch stops (see Figure 15-1). Relative to A, the light travels at the constant speed of 3×10^8 m/sec from ceiling to floor; therefore 10 nsec elapse on his watch, since the distance d traveled is 3 m:

$$t = \frac{d}{c} = \frac{3 \text{ m}}{3 \times 10^8} = 10^{-8}/\text{sec}$$

$$= 10 \times 10^{-9} \text{ sec} = 10 \text{ nsec}$$

Now consider the same experiment as seen by an observer A' who is outside and moving to the left past the room with constant velocity v. According to A', however, he is the one who is "stationary," while the room moves by him to the right with velocity v. At the instant the light flashes, his watch starts and, like A's watch, it also stops when the light hits the detector. When the light flashes, he sees A and the room in the position shown in Figure 15-2 a. A short while later the light is, say, halfway down to the floor and, in the eyes of observer A', the room has moved to the right as shown in Figure 15-2 b. Later still, the light hits the detector and, according to A', the room has moved even farther to the right (see Figure 15-2 c). A' does not observe the light moving straight down from ceiling to floor as A does; to A', the light moves at an angle down and to the right as it travels from the source

15-2

Time Dilation

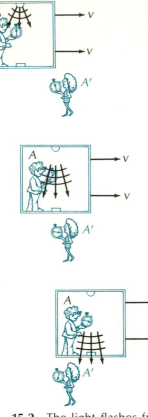

15-2 The light flashes from ceiling to floor as seen by observer A′.

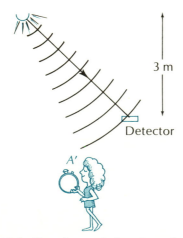

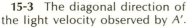

Detector

15-3 The diagonal direction of the light velocity observed by A′.

to the detector (Figure 15-3). This of course is due to the fact that to A′ the detector has been moving to the right during the time the light waves were traveling from the source on the ceiling.

But remember that according to the basic postulate of relativity, A′ measures the light velocity along the diagonal path as $c = 3 \times 10^8$ m/sec, the *same* velocity at which A measured the light as traveling straight down. The diagonal path BD in Figure 15-4 is longer than the vertical path BC; thus observer A′ measures a longer time on his watch for the light to reach the detector than the time stationary observer A registers on his watch in the room. Observer A′ will measure the time as longer than the original 10 nsec by the amount that BD is longer than BC.

The length CD in Figure 15-4 is the distance vt the room travels to the right in time t, the same time that A′ measures the light as traveling from B to D. The length BD is then ct and, by the Pythagorean Theorem, BC is $\sqrt{c^2t^2 - v^2t^2}$. The ratio of BC:BD or $\sqrt{c^2t^2 - v^2t^2}:ct$ is just the familiar square root $\sqrt{1 - v^2/c^2}$. If we call T the time interval of 10 nsec which stationary observer A measures in the room, and T′ the time interval measured by observer A′ who sees the room moving to the right at velocity v, then it is apparent that T is shorter than T′ by the same ratio BC:BD or

$$T' = \frac{T}{\sqrt{1 - v^2/c^2}} \qquad \text{(15-1)}$$

Equation (15-1) states that the time interval measured by A′ on a clock in his reference frame is longer than the time interval measured by observer A by the factor $1/\sqrt{1 - v^2/c^2}$.

Example 15-1:

If stationary observer A's watch registers 10 nsec in the room described above, how long will the same light take by A′'s watch if the room is moving past him at a velocity $v = 0.6\ c = 1.8 \times 10^8$ m/sec? Use Equation (15-1) with $v/c = 0.6$. Then $(v/c)^2 = 0.36$ and $(1 - v^2/c^2) = 0.64$. The square root is $\sqrt{1 - v^2/c^2} = \sqrt{0.64} = 0.8$. Thus the time by A′'s watch is

$$T' = \frac{10 \text{ nsec}}{0.8} = \textbf{12.5 nsec}$$

Thus, if stationary observer A measures "correct" time by a clock at rest with respect to himself, the time recorded by the clocks of any A′ observers moving past A is *dilated*; i.e., time is measured as slower by the factor $\sqrt{1 - v^2/c^2}$ in moving systems.

Remember, however, that this time differentiation is relative. According to observer A', he is stationary and A is moving past him; thus A' measures the watch belonging to A as moving slower by the same amount. How can these apparently contradictory statements both be true? Because time is a relative quantity which is dependent upon the motion of the frame of reference in which it is measured. The old, common-sense idea of an absolute time has to be discarded. Attempts to describe nature in terms of absolute time only produce results that disagree with experiment. We give the name **proper time** to the time interval measured on a clock at rest with respect to the observer; however, this is *not* to be considered as *absolute* time. Many, many tests of the time dilation effect have all borne out the relation in Equation (15-1). No known test disagrees with the prediction of Einstein's special relativity.

If time intervals are measured differently depending on the velocity of the timepiece, then we might suspect that other types of measurements such as length are also relative quantities. And, indeed, lengths do depend on velocity. Consider an experiment which has often been used to verify the time dilation effect. The radioactive nucleus is a particle that decays (changes) into a different type of particle in a certain average time (called the half life) when at rest. When several of these particles move rapidly past us, traversing a given distance L in our laboratory, we measure the rate at which they decay and find that they live a *longer* time by our laboratory clock when they are moving than they do when they are at rest in accord with the time dilation effect.

To illustrate this in a clearer way, we concoct a model for this radioactive decay process. In Figure 15-5, a series of alarm clocks are attached to a board which moves past observer A in the laboratory with a velocity v to the right. All of the clocks are initially actuated when they pass point B. The alarm on clock (1) is set to ring 10 nsec (in its own reference frame) from the time it passes B. Clock (2) is then set for 20 nsec, (3) for 30 nsec, and so on (i.e., each clock is set to ring at equal intervals of 10 nsec after the previous clock). Next we fix the length L over which we observe the clocks and the velocity v, so that we measure a time in the laboratory of, say, just over 40 nsec for the clocks to go from B to C in Figure 15-5. Without relativity, we would expect the alarms to ring on clocks (1), (2), (3), and (4) by the time they pass C. But because of the time dilation, time in the moving clock system is going slower than our time in the laboratory. Thus, with a sufficiently large velocity v, less than 40 nsec are observed to have elapsed on the moving clocks as they pass point C and only

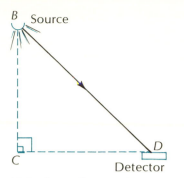

15-4 According to observer A', the light travels from B to D while the detector apparatus travels from C to D.

15-3

Length Contraction

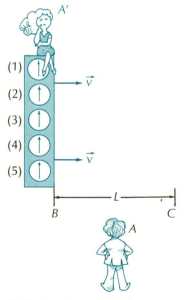

15-5 A series of clocks moves past observer A.

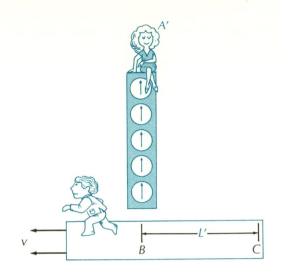

15-6 To observer A', the clocks are at rest while the laboratory moves to the left.

the alarms on clocks (1), (2), and (3) are ringing at that point. The amount of dilation or slowing down of the clocks' time is, as we have seen, proportional to $\sqrt{1 - v^2/c^2}$.

Look at the same experiment from the point of view of an observer A' traveling with the clocks. According to him, he and the clocks are stationary, while the laboratory BC moves past him to the left with velocity v (see Figure 15-6). The instant the clocks pass point B, all of them start and run at their proper rates. But by the time they pass point C, A' hears exactly what the laboratory observer A heard—only clocks (1), (2), and (3) ringing. Since A''s clocks are "correct," he can only conclude that the length L' he measures while the laboratory moves by him must not be the same as the length L measured by the stationary observer A in the laboratory. In fact, L' must be shorter than L by exactly the same factor found for time, $\sqrt{1 - v^2/c^2}$ or

$$L' = L\sqrt{1 - v^2/c^2} \qquad \text{(15-2)}$$

Equation (15-2) states that the space interval (or length in the direction of the velocity) measured on an object moving past an observer with velocity v is shorter than the length when that object is at rest.

Example 15-2:

The length of a rocket at rest is 100 m. If it moves past you at $v = 0.8\ c = 2.4 \times 10^8$ m/sec, what length do you measure? Use Equation (15-2) with $L = 100$ m and $v/c = 0.8$. Then $(v/c)^2 = 0.64$ and $(1 - v^2/c^2) = 0.36$. The square root is $\sqrt{1 - v^2/c^2} = \sqrt{0.36} = 0.6$. Thus the length of the rocket is

$$L' = 100 \text{ m} \times 0.6 = \mathbf{60\ m}$$

Again this measurement of length contraction is relative. We measure the length of the rocket as shortened in the direction it is moving. A person *on* the rocket measures his distances as normal, but distances in our stationary reference frame (which he sees as moving past the rocket) as shortened. So we conclude that distance and space intervals, like time, are not absolute, but depend upon the velocity of the frame of reference in which they are measured. We give the name **proper length** to a length measured at rest with respect to an observer. As in the case of time, the proper length is not to be considered *absolute* length.

Measurements of length and time affect other physical quantities, in particular the measurement of velocities in reference frames moving past us. Suppose that system A' is moving past an observer A with velocity v and an object moves in A' with a velocity u (as measured by A'). Thus, u is the distance x' divided by the time t' as measured by A'. How does observer A measure this?

Consider, for example, a man A' on a moving train (velocity v) throwing a ball with velocity u in the direction of his motion (see Figure 15-7). With the Galilean transformation, the ball thrown with a velocity u by a person A' on a moving train would have a velocity, as seen by observer A on the ground, of just $u + v$, where v is the velocity of the train (of A' with respect to A). But we know that observer A does not measure the same velocity u as observer A' on the train does. The distance x the ball travels is shorter according to A than the distance x' measured by A'. Also the time of flight t of the ball measured by A is longer than t' measured by A'.

The derivation of the velocity addition formula is easily obtained from the Lorentz transformations in Section 14-7, as shown in Appendix 15 (page 305). The result is

$$\text{sum of velocities} = \frac{u + v}{1 + uv/c^2} \tag{15-3}$$

Note that Equation (15-3) gives the classical Galilean result $(u + v)$ when the velocities are small so that the factor uv/c^2 can be neglected. For large velocities u or v, the denominator is larger than one and the sum of velocities will always be less than $(u + v)$.

15-7 Observer A' throws a ball with velocity u as his frame moves past A with velocity v.

Example 15-3:

Rocket A approaches you from the left at a velocity of $0.9\ c = 2.7 \times 10^8$ m/sec. Rocket B approaches you from the right, also at a velocity of $0.9\ c$. What velocity does A measure B as having; i.e., what is the velocity of B relative to A? By Galilean physics, the answer would be the sum of the two velocities that you measure or $0.9\ c + 0.9\ c = 1.8\ c = 5.4 \times 10^8$ m/sec. But that answer is *incorrect*. The velocity addition formula gives

$$\frac{0.9\ c + 0.9\ c}{1 + 0.9 \times 0.9} = \frac{1.8\ c}{1.81\ c} = \textbf{0.995}\ c$$

Thus the speed of B measured by A is still less than the velocity of light c.

Example 15-4:

A rocket is approaching you with a velocity $v = 0.9\ c$. A light on the rocket flashes on and is emitted from the ship at c, the speed of light. With what velocity do you measure the light? Again, use Equation (15-3), with $v = 0.9\ c$ and $u = c$. The velocity sum is

$$\frac{0.9\ c + c}{1 + (0.9\ c \times c/c^2)} = \frac{1.9\ c}{1.9} = c$$

The velocity you measure for light is exactly c, and this result is in agreement with the basic postulate of relativity: **The velocity of light is always the same**; it does not change if the source is moving.

The velocity addition result is disturbing to most people who encounter relativity for the first time. This is because our common-sense ideas about the addition of velocities are violated. But we have to accept the fact that these ideas are wrong in that they do not agree with experiment, with nature, whereas the relativistic explanation represented by Equation (15-3) does.

15-5

Momentum and Mass Increase

The measurements of lengths and times and velocities for moving objects are not in accord with our common-sense ideas of space and time. Remember, however, that all the results of the Mechanical Revolution—Newton's laws, energy, momentum, and so on—used these same common-sense concepts of space and time. On philosophical grounds, all classical Newtonian physics must be discarded, since the universe we measure does not agree with the Newtonian explanation. But, as we will see later, when the velocity of an object is small in comparison with the velocity of light, Newtonian mechanics can approximate the new picture relativity gives us very closely.

Einstein proposed that when two particles interact with one another, the principles of conservation of momentum and conservation of energy should hold for *all* observers moving at arbitrary velocities with respect to the two interacting particles. Energy and momentum were thus singled out to occupy a very basic role in relativity. Newton had based his laws on the forces between objects and the accelerations those forces produced, but, as we have already seen in Chapter 7, the energy and momentum methods allow us to solve problems without using forces and consequently provide an alternative to the force concept. In particular, the use of energy and momentum is essential in problems for which the forces are unknown. With relativity, the distinction between the two rival concepts—forces and accelerations on one hand and energy and momentum on the other—becomes even greater.

Einstein's hypothesis on the universality of the principle of conservation of momentum, no matter what the state of relative motion of the observer, has been experimentally verified over and over again and no violation of it has ever been observed. We regard it as providing a "correct" description of nature in the same way that the earlier hypotheses of Copernicus, Galileo, Newton, and Maxwell were once regarded as "correct."

By postulating momentum conservation to be true for all observers, Einstein arrived at a new discovery, a new concept of mass. He found that the mass of an object was not measured as the same by observers with different velocities relative to the object: The greater the velocity of the observer relative to the object, the larger the mass he measures for the object. To see why this is true, imagine that you throw a putty ball of mass m with a velocity u at a cart of mass M (Figure 15-8 a) as in Example 7-2 (page 95). The putty will stick to the cart and, because momentum is conserved in the direction of the velocity u (call this the x direction), the cart and the putty will both recoil at a velocity $U = [m/(M + m)]u$ (Figure 15-8 b).

If the cart is constrained so that it can only roll along the track in the x direction and the entire track is moving at a velocity of v in the y direction (perpendicular to the x direction as shown in Figure 15-9), the putty will still have the same momentum mu in the x direction and the x component of the momentum after the putty hits and sticks to the cart will therefore be the same as well, regardless of the fact that the cart is also moving in the y direction. This is because momentum is a vector quantity and the x component is independent of any motion in the y direction. So the cart still moves down the track in the x direction with velocity $U = [m/(M + m)]u$.

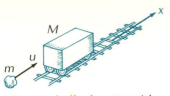

15-8a A ball of putty with velocity u about to hit the cart.

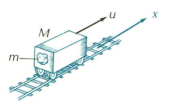

15-8b Conservation of momentum shows the velocity of the cart and putty after collision is $U = [m/(M + m)]u$.

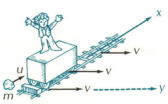

15-9 The entire cart and its track move with velocity v in the y direction.

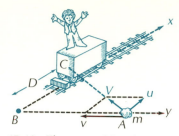

15-10 The putty hits the cart from the point of view of an observer on the cart.

Finally, we look at this same experiment from the point of view of an observer on this constrained cart. To him, the putty is approaching at an angle with a velocity u along the x axis and a velocity v in the y direction. The vector sum of these two velocities V is pointed directly at the cart (as in Figure 15-10). Since the cart can only recoil in the x direction, we would find experimentally that it has the same velocity U down the track after collision. In other words, the conservation of the momentum in the x direction is not affected by any velocity the cart or the putty may have in the y direction.

Assume now that u is small compared with c (the velocity of light), so that lengths and times due to the x component of motion are not affected appreciably ($\sqrt{1 - u^2/c^2} \approx 1$). However, v may be large. From the point of view of the observer on the cart, the velocity of the putty along the x axis is measured as the distance D along the x axis (Figure 15-10) divided by the time T the putty takes to go from A to C (or from A to B for the component of its motion along the y axis). Since u is small, there is no significant change in D due to the x motion. But, assuming v is not small, the stationary observer on the cart will measure a different time T for the putty to go from A to C in Figure 15-10 than the time used in the reference frame of the putty (Figure 15-9). In fact, due to time dilation, the observer on the cart will measure a time interval T which is larger by the factor $1/\sqrt{1 - v^2/c^2}$ than the time used in the reference frame of the putty. Since the velocity in the x direction is D/T, the cart observer measures an x velocity of the putty as smaller by the same factor $\sqrt{1 - v^2/c^2}$ than the velocity u which the putty had in its own reference frame. Thus, the cart observer would measure a smaller x momentum $mu\sqrt{1 - v^2/c^2}$ than the x momentum mu measured in the putty's reference frame (Figure 15-9).

Yet the momentum of the putty cannot be smaller; it must still equal mu since the cart still moves down the track with the same velocity U as before. Einstein resolved this apparent contradiction when he assumed that the mass m' of the putty measured by the observer on the cart is larger than the mass m of the putty at rest (or moving slowly). If the mass m' is measured as larger by the factor $1/\sqrt{1 - v^2/c^2}$ due to its motion with velocity v, then the x component of momentum will be the same before and after collision. Thus an object moving past you with velocity v has a mass m' which is larger than its **rest mass m** when it is stationary, or

$$m' = \frac{m}{\sqrt{1 - v^2/c^2}}$$ (15-4)

Example 15-5:

A one-gram cube of matter measured at rest moves by you with a velocity $v = 2.4 \times 10^{10}$ cm/sec. What mass do you measure for the cube? Use $m' = m/\sqrt{1 - v^2/c^2}$, with $m = 1\,\text{g}$, $v = 2.4 \times 10^{10}$ cm/sec, and $c = 3 \times 10^{10}$ cm/sec. Then

$$\frac{v}{c} = \frac{2.4 \times 10^{10}}{3.0 \times 10^{10}} = \frac{4}{5}$$

and

$$\frac{v^2}{c^2} = \frac{16}{25}$$

The square root of this is

$$\sqrt{1 - \frac{v^2}{c^2}} = \sqrt{1 - \frac{16}{25}} = \sqrt{\frac{9}{25}} = \frac{3}{5}$$

The observed mass is then

$$m' = \frac{1\,\text{g}}{3/5} = \frac{5}{3}\text{g} = \textbf{1.67 g}$$

Let's try to summarize the above argument. The principle of conservation of momentum holds independently for either of the space directions x or y. For momentum to continue to be conserved in the x direction, therefore, any motion in the y direction must not affect the amount of momentum $= mu = m(D/T)$ along the x axis. But we already know that the time T is measured as larger (the time dilation effect) due to the object's motion along the y axis. Thus, in order that the x momentum $m(D/T)$, be the same, the mass we measure must increase by the same amount as the time increases, so that the increase in time in the denominator is exactly cancelled by the increase in mass in the numerator, leaving the momentum unchanged. As in the cases of time and length, the mass increase has been verified experimentally many times and always agrees with Einstein's formula, Equation (15-4).

The second of Einstein's two famous relativity papers of 1905 was entitled "Does the inertia [mass] of a body depend on its energy content?." Although this is a novel and far-reaching question, we will see that it follows directly from the mass increase just discussed in the last section. In Chapter 7, we learned that kinetic energy is a property that a body has when it is moving with velocity v ($KE = \frac{1}{2}mv^2$), and that the larger the velocity, the larger

15-6

Energy and Mass

the energy. But according to the results in Section 15-5, when an object has a velocity with respect to you, its mass increases. There must therefore be a direct relation between mass and kinetic energy. Einstein showed that this relation would be

$$KE = m'c^2 - mc^2$$
$$= (m' - m)c^2 \qquad (15\text{-}5)$$

where KE is the kinetic energy, m is the mass of the object at rest, m' is the mass of the moving object, and c is the velocity of light. (The proof of this relation is beyond the mathematical level of this book, since it requires a knowledge of the calculus.)

To prove that Equation (15-5) indeed does represent the kinetic energy, Einstein showed that for small velocities, for which Newtonian physics should be nearly valid, Equation (15-5) took the form

$$\tfrac{1}{2}mv^2 = (m' - m)c^2 \qquad (15\text{-}6)$$

which is the Newtonian result for kinetic energy. This is only the limiting case of Equation (15-5) when v is much smaller than c.

If $(m' - m)c^2$, expressed as $m'c^2 - mc^2$ without parentheses, is the kinetic energy due to motion, then Einstein argued that each term must separately represent an energy. From this he interpreted $E' = m'c^2$ as the **total energy** of an object moving past an observer and $E = mc^2$ as the **rest energy** of an object when it is not moving. The difference between E' and E is then the kinetic energy given in Equation (15-5). The equality of mass and energy

$$E = mc^2 \qquad (15\text{-}7)$$

shows that any object which has mass has an inherent energy or, conversely, anything that has energy (like light) must also possess mass.

One new concept we have to accept is that the idea of potential energy PE has disappeared, replaced by **mass energy** mc^2. The total energy used to be composed of kinetic energy KE plus potential energy PE:

$$E_{\text{total}} = KE + PE$$

Now the total energy according to relativity is kinetic energy KE plus mass energy

$$E_{\text{total}} = KE + mc^2 \qquad (15\text{-}8)$$

For example, an object of mass m at rest at height h above the surface of the ground was considered to have potential energy equal to mgh capable of changing into kinetic energy of motion if the object were dropped (Section 7-5, page 98). With Einstein, this concept changed in that the combined mass of the (earth +

object) became the mass energy. When an object is at a height h above the earth, the total combined mass of the object and earth is *greater* than it is after the object has fallen to the ground. Thus the total mass changes; some of the mass energy is converted into kinetic energy when the object is dropped. We will examine this concept in greater detail when we discuss atomic and nuclear physics later.

Example 15-6:

What is the total energy available in 1 g of matter, if all of the mass could be converted into energy? Use $E = mc^2$ with $m = 1$ g and $c = 3 \times 10^{10}$ cm/sec.

$$E = 1\,g \times (3 \times 10^{10}\ \text{cm/sec})^2 = \mathbf{9 \times 10^{20}\ ergs}$$

This is an incredibly large amount of energy. It is about the same as the kinetic energy a freight train weighing 100 tons would have when traveling at 50 km/sec or about 30 mi/sec.

Because the mass of an object increases with velocity, it is apparent that any material object could not be speeded up to a velocity equal to or greater than the velocity of light c. Because of $E = m'c^2 = mc^2/\sqrt{1 - v^2/c^2}$, it would require an infinite amount of energy for a particle to reach the velocity $v = c$, and an infinite amount of energy is not available. However, elementary particles obtained from high-energy particle accelerators have been accelerated to within a fraction of a percent of the velocity of light, and under such conditions the energy relations in this section hold exactly.

Questions

1 Are the relativistic phenomena of length contraction, time dilation, mass increase, etc., involved with "actual" changes in length, time intervals, mass, etc., or are they only involved with apparent changes? Explain.

2 Explain how predictions can be made on the basis of special relativity which violate common sense, yet are reliably established by experiment.

3 Discuss some of the phenomena which you might observe if the speed of light were only 20 mi/hr.

4 Discuss what you might observe if the speed of light were actually infinite. What would happen to the transformation equations of Lorentz and Einstein?

5 What would happen to the length contraction, time dilation, and mass increase if the speed of light were infinite?

1   The formula $E = mc^2$ seemingly contradicts the laws of both conservation of mass and conservation of energy. Explain how this
equation affects these laws and what new conservation law results
from it.

2   Is there experimental evidence to substantiate the physical predictions of relativity? Research and discuss some of the experimental
evidence for the phenomena of length contraction, time dilation, and
energy-mass equivalence ($E = mc^2$).

3   Explain why the "clock paradox" does not actually occur.

4   Find experimental evidence for the relativistic mass increase. Do you
think this evidence is conclusive or do you believe there still might
be some other explanation for the increase?

Problems

1   A spaceship with a proper length of 100 m is moving at a speed of
0.8 c with respect to an observer. What does the observer measure
the ship's length to be?

2   If the spaceship in Problem 1 is going so fast that its measured length
is 60 m, what is its speed?

3   A light on the spaceship in Problem 1 is timed to flash every 10 sec.
What interval between flashes will a stationary observer record?

4   What is the formula for the relativistic "addition" of velocities? Use
this formula to show that if one of the velocities is c in one frame,
the velocity is c in every frame.

5   If the spaceship in Problem 1 has a rest mass of 20,000 metric tons
(2×10^{10} g), what mass does the stationary observer see?

6   Use the formula $E = mc^2$ to find the energy contained in 1 kg (2.2
lb) of matter. Compare this with the electric energy generated each
day for some large city.

7   Observers A and B are on two spaceships which are approaching
the earth from opposite directions. The speed of A (relative to the
earth) is 0.9 c and that of B is 0.4 c. With what speed does A observe
B approaching him? With what speed does B observe A approaching?

8   A certain star is receding from the earth at a speed of 0.1 c. If the
star emits light of wavelength $\lambda = 2.000 \times 10^{-8}$ cm, what do we
measure as its wavelength?

9   The star Alpha Centauri is 4 light years from earth (measured from
the earth). A spaceship travels at a speed of 0.9 c between the earth
and Alpha Centauri and back again. How long does the trip take
as measured by clocks on the earth? By clocks on the spaceship?

10 What distance separates the earth and Alpha Centauri to an observer
on the spaceship in Problem 9?

11 Show that it is possible for two events to occur in different time
order in different systems. (*Hint:* Start with a system in which two
events occur at the same time coordinate, but at different space
coordinates. Then look at the order of the events in two other frames
moving in opposite directions.)

Other Effects of Relativity

16

The main results of special relativity derived and discussed in Chapter 15—the time dilation, length contraction, velocity addition, mass increase, and energy-mass formulas—all follow logically and consistently from the fundamental postulate that the speed of light is independent of any (constant velocity) motion of the observer or of the source. Although most of the physical effects (the dilation of time, the shortening of lengths, and so on) are outside our everyday sense experiences, they have all been experimentally verified time and time again. Einstein's special relativity is now considered a "law" rather than a "theory." Today, physicists incorporate special relativity as a matter of fact into their theories of elementary particles.

However, several loose ends still exist which we will clear up in this chapter. One of these—the relativistic relationships between energy and momentum—will prove useful later in understanding Chapter 20 on particle physics. And, as we will soon see, Einstein's theory of relativity also solves the apparent dilemma mentioned at the beginning of Chapter 14—that the electrodynamic force between two charges is different when measured in a reference frame where the charges are moving—by using the results of the length contraction effect. Finally, this chapter concludes with a brief discussion of the main ideas of Einstein's Theory of General Relativity which deals with *accelerated objects* (in contrast to his Theory of Special Relativity which deals with objects moving at a constant velocity).

16-1

Introduction

16-2

Relativistic Energy-Momentum Relation

The momentum of a particle depends upon both its mass and its velocity. Its kinetic energy also depends on the same two quantities and, as a consequence, one can derive an expression that directly connects the energy and the momentum. In classical Newtonian physics, the kinetic energy $KE = \frac{1}{2}mv^2$ is related to the momentum $p = mv$ by the equation $\frac{1}{2}mv^2 = p^2/2m$. We note that the first power of the energy KE is related to the square (second power) of the momentum p. In relativity physics, the total energy $E = m'c^2$ is related to the momentum $p = m'v$ by making use of the mass increase formula (Equation (15-4)):

$$m' = \frac{m}{\sqrt{1 - v^2/c^2}}$$

and a little algebra, shown in Appendix 16 (page 306), to obtain

$$E^2 - p^2c^2 = m^2c^4 \qquad (16\text{-}1)$$

Several things can be noted about Equation (16-1). First, unlike the classical formula, it relates the square (second power) of the energy to the square of the momentum; (this will make it important later when we discuss particle physics). Secondly, the term m^2c^4 on the right-hand side of the equation is a constant, since m is the *rest* mass which always has the same value and c is the constant speed of light. Therefore Equation (16-1) should be true for *all* reference frames, no matter what velocity exists. An observer A' moving past you with velocity v measures an energy E' and a momentum p' for some object. Although you measure different values of E and p, for A' Equation (16-1) also holds: $E'^2 - p'^2c^2 = m^2c^4$. Since the quantity $E^2 - p^2c^2$ in Equation (16-1) equals the same constant (m^2c^4) as $E'^2 - p'^2c^2$ above, the two quantities must also be equal:

$$p^2c^2 - E^2 = p'^2c^2 - E'^2 \qquad (16\text{-}2)$$

This equation relates the total energy E and the total momentum p in one reference frame to the total energy E' and the total momentum p' in a second reference frame which is moving relative to the first.

Equation (16-2) is strikingly similar to the basic space-time equation of Einstein's relativity [(14-5), page 185]. In fact, they are identical if we replace pc with x, $p'c$ with x', E with ct, and E' with ct'. This information can now be used to find the solutions to Equation (16-2)—the transformation equations for E and p in terms of E' and p'. The Lorentz transformations of Equations (14-6)–(14-9) were the solutions of Equation (14-5); the same equations, with the substitutions just mentioned must therefore satisfy Equation (16-2).

Look again at the structure of the Lorentz transformation equations (14-6 and 14-7) for space and time:

$$x = \frac{x' + vt'}{\sqrt{1 - v^2/c^2}} \qquad (14\text{-}6)$$

$$t = \frac{t' + vx'/c^2}{\sqrt{1 - v^2/c^2}} \qquad (14\text{-}7)$$

One striking property of these equations is that they inextricably mix space and time together. When we transform from one coordinate frame to another frame which is moving relative to the first, a length x in the first frame is found only in terms of both x' and t' in the second; likewise, x' and t' are both needed to find t. So time and space can no longer be considered separate physical entities. Relativity tells us that the intervals in **space-time** are more basic than separate time intervals and separate space intervals.

An event E_1 occurs at time t_1 and position x_1 and a second event E_2 occurs at time t_2 and position x_2. Although we can still talk in terms of the time interval $(t_2 - t_1)$ between the two events, this is not a particularly useful concept since, as we discovered from the time dilation effect, the time interval is not constant but is measured differently by observers in different states of relative motion. Likewise, the concept of the space interval or length $(x_2 - x_1)$ is not as useful by itself because it is also dependent upon the state of relative motion (the length contraction effect). But a *space-time* interval $(s_2 - s_1)$ can be defined as the same in all reference frames if s_1 is interpreted as

$$c^2 t_1^2 - x_1^2 = s_1^2$$

so that the space-time interval $(s_2 - s_1)$ is

$$c^2(t_2 - t_1)^2 - (x_2 - x_1)^2 = (s_2 - s_1)^2$$

Note that, for light, $c^2 t^2 - x^2 = 0$ and the space-time interval $(s_2 - s_1)$ is always zero. For particles, which have a *rest mass*, $(s_2 - s_1)$ is *nonzero*.

Now refer back to Equation (14-5), the basic equation of relativity (page 185), and note that the space-time "location" s of an event has the same description in every reference frame. Therefore, the space-time interval $(s_2 - s_1)$ is the same—is unchanged in all coordinate frames *regardless of their relative motion*. Because this space-time interval $(s_2 - s_1)$ between two events remains unchanged from one frame to another, it is considered to be more fundamental in nature than space alone or time alone.

Moreover, since time and space are now linked together in the manner just described, the time t, or rather the time multiplied by the velocity of light ct, appears formally in all relativistic equations as another dimension, like the x, y, and z of three-dimensional space. Thus the space-time description of events, which appears to be nature's simplest description, is inherently *four*-dimensional. We, as observers, separate these four dimensions into the familiar three dimensions of space x, y, and z and the one dimension of time *ct*. It appears to be extraordinarily difficult for most people to think or to live in terms of space-time, perhaps because all our lives we have been used to thinking in terms of separate space and separate time.

The same arguments exist for momentum and energy. They are also linked together by relativistic transformations so that, to the three components of momentum x, y, and z, we can add a fourth dimension or component—the energy (actually E/c). Physicists speak of momentum-energy as a four-vector or a vector in a four-dimensional space-time.

16-4
Ampère's Law for Current-Current Forces

One extraordinary result of the length contraction effect is that Ampère's law for the force on two currents, or the force on two moving charges, can be *derived* from Coulomb's law for stationary charges. This means that, in reality, Coulomb's law is the *only* force law for all electromagnetic forces. Ampère's law can no longer be considered a basic force law, but simply an "apparent" force due to the relative motion of charges. We will now show qualitatively why the current-current force arises.

Consider the two parallel wires A and B in Figure 16-1. Each wire has an equal number of freely-moving negative charges (*electrons*) and heavy positive ions, fixed in the crystalline structure of the metal. Because of the equal number of positive and negative charges, each wire is electrically neutral and no electrical force exists between the two wires.

Now we connect both wires to a battery so that a current flows in the same direction in each wire. The positive ions in both wires remain stationary and the electrons all move to the right with a drift velocity *v* (Figure 16-2). In any length, say one inch, of either

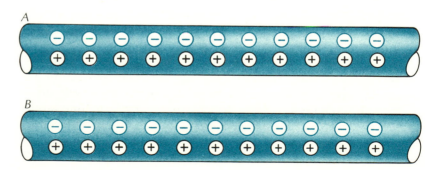

A

B

16-1 Two parallel wires, each electrically neutral.

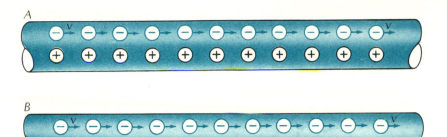

wire there are a number n positive ions and n electrons. When the electrons are at rest, there is no net force. But once the electrons move, there is a net force. Imagine we are moving with the electrons and they are at rest relative to us, so we can apply Coulomb's law to them. The number of electrons per inch n' in this "electron frame" is less than n in the laboratory frame because of the length contraction. The same number of electrons (at rest) must be contained in a longer distance in the electron frame than when moving in the laboratory. With fewer electrons per inch n', the electron-electron force of repulsion is less than the electron-positive ion force of attraction, and we observe that Coulomb's law predicts a net attractive force between wires A and B—the actual interaction between two wires carrying currents in the same direction.

And so all of the "laws" for moving charges can be derived exactly, simply by the use of Coulomb's law and the results of Einstein's special relativity. Thus magnetism (except perhaps for quantum effects), which arises from currents or the motion of charges, is a relativistic effect—a rather striking example of how the length contraction effect is "observable" in our everyday experience. Every time you run an electric motor, the force that makes it operable is a current-current force arising from the length contraction. Although the amount of the length contraction is very, very small due to the small value of the drift velocity, the force is still large due to the extraordinarily large number of electrons and positive ions present in the wire.

Relativity tells us that all physical quantities that we measure—space, time, mass, energy, momentum, and force—are relative. They do not have any absolute, "true" values, only the values we measure—values which change from one reference frame to another. When the objects we measure are at rest with respect to

16-5

The Role of the Observer

us, we speak of proper length, proper time, rest mass, but we cannot consider these to be "true" values because we will measure different lengths, times, and masses when these objects are moving which will be just as "true" as our stationary measurement values.

The concepts of absolute length, time, and mass must therefore be discarded. We, as the observers, can no longer measure absolute physical quantities belonging to an object "out there" in space. We can no longer say that the observer has a passive, unimportant role in what is observed. Relativity shows us, on the contrary, that the state of motion of the observer relative to the object is critical in determining all of the physical quantities relating to that object. This is something new in physics and in science in general. Most scientists previously held that the object being observed in itself possessed absolute physical quantities like length, mass, and time which an observer merely observed—that the observer himself could not change the "absolute" quantities of an object. But now we see that, simply by moving relative to the object, an observer can and does change these quantities: He measures them differently, hence they are not absolute. Physical quantities are partly dependent on the object itself and partly dependent on the observer's state of motion relative to that object.

With relativity, the observer begins to take an active role in determining the physical properties of his own external reality. We will see in the following chapter that quantum mechanics reveals a similar fact—that the observer is not detached, but is an integral part of what he observes. Both of these new insights into the observer's role seem to be leading to something which is as yet unclear. For now, perhaps the most we can say is that physics is entering a new conceptual phase, one in which the observer himself (the measuring apparatus) plays an important part in the external measurement of the world around him.

16-6

General Relativity

About six years after the publication of his Special Theory of Relativity in 1905, Einstein published the first of a long series of papers in which he tried to generalize his theory to accommodate accelerated motion. First he connected accelerations with gravity and, as a result, geometrized the gravitational "force." Most of the remainder of his physics career was spent in an unsuccessful attempt to unify the gravitational and electrical fields.

The "Principle of Equivalence" marked the beginning of Einstein's Theory of General Relativity. Imagine you are in a windowless room on the surface of the earth. You try various experiments to determine that there is an acceleration of gravity g equal to 32 ft/sec^2. You test your weight on a set of scales, drop a ball and time its fall down to the floor, time the period of oscillation of a pendulum, and so on (Figure 16-3). Now suppose you are in this same room far out in space away from all gravitational fields. The entire room and its contents are accelerated upward in the direction of the ceiling at 32 ft/sec². Einstein's equivalence principle states that the results of all experiments in the accelerated room will be the same as they were in the stationary room sitting in a gravitational field of the same amount (see Figure 16-4): There is no way to distinguish between the acceleration of a gravitational field and the acceleration due to increase in motion.

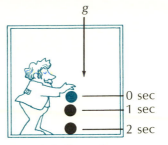

16-3 A ball accelerates downward in a gravitational field g.

Recall that, in Chapter 4, Copernicus, Galileo, and Descartes wrestled with the principle of inertia which evolved slowly during the sixteenth and seventeenth centuries. The inertial mass of a body was later defined by Newton as the resistance that body offered to accelerations. However the gravitational mass which determined the force on the body in a gravitational field was found experimentally to be equal to the inertial mass. This is just the equality which Einstein reinvestigated in his General Theory. In his own words,

> *The equality of these two masses, so differently defined, is a fact which is confirmed by experiments of very high accuracy . . . and classical mechanics offers no explanation for this equality. It is, however, clear that science is fully justified in assigning such a numerical equality only after this numerical equality is reduced to an equality of the real nature of the two concepts. . . . A little reflection will show that the law of the equality of the inert and the gravitational mass is equivalent to the assertion that the acceleration imparted to a body by a gravitational field is independent of the nature of the body. For Newton's equation of motion in a gravitational field, written out in full, is*

$$(Inert\ mass) \cdot (Acceleration) = (Intensity\ of\ the\ gravitational\ field) \cdot (Gravitational\ mass)$$

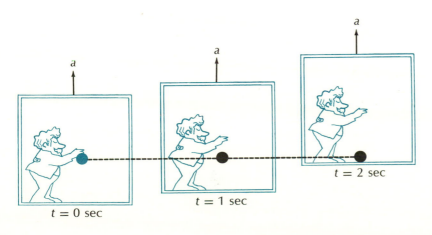

$t = 0$ sec

$t = 1$ sec

$t = 2$ sec

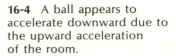

16-4 A ball appears to accelerate downward due to the upward acceleration of the room.

It is only when there is numerical equality between the inert and gravitational mass that the acceleration is independent of the nature of the body.[1]

Einstein went on to describe two systems of coordinates, one "at rest" with a gravitational field and the other with no gravity, but accelerating uniformly with respect to the first:

The assumption of the complete physical equivalence of the systems of coordinates . . . we call the "principle of equivalence"; this principle is evidently intimately connected with the law of the equality between the inert and the gravitational mass, and signifies an extension of the principle of relativity to coordinate systems which are in non-uniform motion relatively to each other. In fact, through this conception we arrive at the unity of the nature of inertia and gravitation.[1]

At this point, Einstein's Principle of Equivalence was merely an hypothesis, but he used it to make predictions of nature which he then tested. The first prediction arose from the obvious fact that, in a room accelerating upward, a horizontal light beam from a flashlight will strike the opposite wall slightly below the spot where the light is aimed. This is easy to verify, since in the time it takes the light to travel from the flashlight to the wall, the entire room will have accelerated upward slightly. So the person in the room, who previously measured his usual earth weight and saw the ball accelerate "down" as it would in a gravitational field, will now measure the light as "bending" down as it traverses the room. From the equivalence principle, the light is also bent by a gravitational field. Striking confirmation of this prediction for starlight passing close to the sun (a large gravitational field) was first made in 1919 when a star, which would not normally have been visible at a particular instant when it was "behind" the sun, was observed because its light rays were bent by the sun's mass as they traveled past (see Figure 16-5). This effect has been confirmed many, many times since then during total eclipses of the sun. (An eclipse is necessary in order to block out the intensely bright disc of the sun so that the faint starlight can be seen.)

[1] Albert Einstein, *The Meaning of Relativity,* p. 57. Copyright 1953, Princeton University Press; 1956, estate of A. Einstein. Used by permission of Princeton University Press.

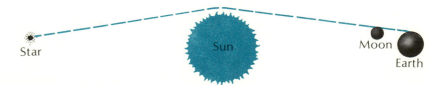

16-5 The bending of starlight by the sun's mass.

Star Sun Moon Earth

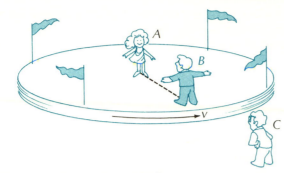

16-6 Observers on a merry-go-round.

Further results of general relativity can be qualitatively described using a simple model attributable to Einstein. Imagine a merry-go-round with an infinitely thin observer A at the very center, an observer B at its edge, and an observer C off the merry-go-round at rest on the ground. A and C are in the same inertial or reference frame since they are not moving with respect to one another. (If the rotation of A bothers you, let A be a very thin pole with many sensing and measuring devices fixed on all of its sides.) Therefore A and C must agree in their measurements of length, time, mass, and so on. However, B is moving past C with a velocity v, the peripheral or tangential velocity of the merry-go-round due to its rotation (see Figure 16-6). Therefore C measures B's time as slower and B's lengths as shorter in the direction of the motion (along the circumference of the merry-go-round), but the length of the radius from A to B is not shortened since it is perpendicular to the velocity. It is clear A must agree with C that B's time is slower than either A's or C's and that the circumference of the merry-go-round has been shortened.

But how can this be? If there were no outside reference points—if the entire merry-go-round were enveloped by complete darkness—A would not even be aware of the rotation. A measures B's time as slower, yet B is not moving relative to A. If A walks out to B to find out why, he finds something he did not experience at the center—an outward force, the centrifugal force you experience on any rotating object. Therefore A concludes that whenever there is a force—or acceleration, by Newton's Second Law—time goes slower and lengths perpendicular to the force contract. This should be true of both acceleration due to changes in motion and of gravitational acceleration due to the Principle of Equivalence.

Apparently these bizarre effects are no longer relative, as they were in special relativity where constant velocities existed. Accelerations and forces can be felt or measured by accelerometers and we know which of two observers is being accelerated as they approach one another. One feels the acceleration and his time is slower; the other does not feel the acceleration and his time does not change.

Because of the equivalence principle, time should (and does) go slower in a large gravitational field (like the sun's) than it does on earth. This has been measured and has confirmed Einstein's predictions. In fact time should be measured as slower on the first floor of a building than on the upper floors of that building where the earth's gravitational field is slightly weaker. (This effect was also measured in an ingenious experiment performed at Harvard University in 1960.)

The fact that the circumference of the merry-go-round gets shorter while its radius does not means that, in the presence of accelerations or gravitational fields, the Euclidian geometry of a flat plane no longer holds. The circumference of the merry-go-round is not equal to $2\pi r$ as it would "normally" be, but is less than $2\pi r$. Although we are talking about four-dimensional space-time here, a two-dimensional analogy will give us an idea of what this means. For the circumference of a "circle" to be less than $2\pi r$, the circle's area cannot be flat but must be curved as on a sphere (Figure 16-7) or inside a bowl. From this, Einstein worked out the rather complicated mathematics of curved or warped space-time, which arises in a strong gravitational field near a large mass. He was successful in that his theory predicted with astonishing accuracy anomalies in the orbit of the planet Mercury (nearest the sun's strong gravitational field) that had been known for years but could not be explained by Newtonian Mechanics.

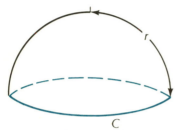

16-7 For the circumference C of a circle to be less than $2\pi r$, the circle's area must be curved.

Questions

1   The interval s in space-time is difficult to define "physically." Is the same true for the analogous interval mc^2 in momentum-energy?

2   How is a four-dimensional vector different from ordinary three-dimensional vectors? How is it similar?

3   In view of the fact that light has no rest mass, what do you think can be meant by the momentum of light?

4   Relativistic effects are not normally observed in our everyday life because the velocities of the objects that we can see are very small compared with the velocity of light. Explain why the force between two currents provides an exception to this statement. (The velocity of electrons in currents is about 0.1 cm/sec.)

5   Can the bending of starlight near the sun be considered a warping or curving of space-time? Explain.

6   Would light be bent by the gravitational field of all the matter in the universe? Explain.

7   Explain how the length contraction effect solves the dilemma that some observers see Ampère's law between currents whereas other observers in different reference frames do not.

1   Would you expect the velocity of light to change in a strong gravitational field? How and why?

2   In an exceedingly strong gravitational field (like a heavy "neutron" star), light might be curved so violently that it would be captured by the star. What would we see of this process from the earth? Explain.

3   What other properties of special relativity besides the time dilation and length contraction effects can be applied in general relativity?

4   Referring back to Question 6, what would happen to light emitted in the universe if the universe were finite (i.e., the universe were like a sphere with a boundary)? If the universe were infinite?

Problems

1   A proton whose rest mass is 1.67×10^{-24} g is accelerated by a cyclotron (a particle accelerator) until its total energy is 2×10^{-3} ergs. What is its momentum?

2   What is the relativistic mass of the proton in Problem 1? What is its velocity?

3   When the kinetic energy of a single particle is zero, what is its velocity? (*Hint:* Check the equality of the momentum-energy relation for this case.)

4   Can the combined momenta of two particles be zero, yet have a nonzero kinetic energy? In what reference frame could this be true?

5   Use the energy-momentum relation to show that for particles with a zero rest mass (light), the energy E is related to the momentum p as $E = pc$.

6   A particle that does have rest mass acquires more and more energy. When this acquired energy becomes very large, show that E is related to p approximately as $E = pc$.

7   Show that for an energy E very near the rest mass (small KE) the classical relation holds between kinetic energy and momentum $(\mathrm{KE} = p^2/2m)$.

8   Use the same argument given in Section 16-4 to show that two currents in opposite directions will repel one another.

17

The Beginning of
Quantum Mechanics

17-1

Introduction

If the Michelson-Morley experiment (Chapter 13) was one of the clouds obscuring "the beauty and clearness of the dynamical theory which asserts light and heat to be modes of motion," another of the clouds to which Lord Kelvin referred in 1900 consisted of several unexplained and seemingly unrelated experiments. The optimistic feeling of the time was that the relevance of these experiments would eventually be clarified by the powerful tools that only Newtonian mechanics and Maxwellian electrodynamics could provide. How wrong this feeling turned out to be. We will not attempt to describe here all of the pre-1900 evidence which indicated that a new revolution in physics was brewing, but we will concentrate on a few of the more important pieces in this chapter.

17-2

Photons—The First Basis of Quantum Mechanics

The first piece of evidence that all was not quite so rosy resulted from experiments that measured the colors emitted from a very hot object. Experimental observations of the wavelengths of electromagnetic radiation emitted from a heated thin shell of metal were found to differ significantly from the predictions of the "best" theory based on classical mechanics and electrodynamics. The theory predicted that the heated metal would radiate an infinite amount of energy of very short wavelength, or high-frequency electromagnetic radiation in the form of ultraviolet light, x rays, or even higher-frequency radiation. If a thin-shell metal cavity were energized with a relatively low-frequency light or infrared radiation, it should, according to theory, soon resonate at all multiple higher frequencies in much the same way that a piano or a guitar string resonates at the higher harmonics above the "fundamental" or lowest note—an effect which physicists of the day referred to as the "ultraviolet catastrophe." However, actual experiments showed that no such catastrophe existed in nature; the heated metal radiated only a finite amount, indeed only a vanishingly small amount, of energy at high frequency—a result which simply could not be explained by the combined power of Newtonian-Maxwellian classical physics. Many physicists sought to find an explanation based on classical concepts, but nature stubbornly refused to fit the old, established ideas. Max Planck's successful interpretation of these experimental results in 1900, while it explained the data with an economy of new constructs, was not immediately accepted because it required a major break with traditional thought to do so.

It had always been assumed that light waves were analogous with water waves, sound waves, and so on, in that their energy was dependent upon amplitude: The brighter the light, the greater the amplitude and the greater the energy; any amount of energy could thus be delivered by electromagnetic radiation or light if it had enough intensity or brightness and were left on for a long enough time. Although this picture conformed to most experimental observations, it predicted an "ultraviolet catastrophe" in this particular case which did not occur in nature.

Planck alternatively proposed that, when light was emitted from a radiating object like the metal cavity, the energy of the light was proportional not to the *amplitude* of the wave, but rather to the *frequency* of the wave. He then wrote an equation for the energy E of electromagnetic radiation

$$E = hf \tag{17-1}$$

where f is the frequency of the wave in vibrations/sec and h is a constant ($h = 6.6 \times 10^{-27}$ erg-sec) now known as Planck's constant. This new property of electromagnetic radiation was meant to sustain the **emission** of radiation; presumably the radiation would then behave like a normal electromagnetic wave does after leaving its source.

It is fairly easy to understand why this new idea of energy, Equation (17-1), resolved the dilemma of the ultraviolet catastrophe. If a certain finite amount of heat or light energy E_0 were originally put into the cavity, then the higher-frequency electromagnetic oscillations would try to share this energy as before. If, however, the higher the frequency, the larger the energy ($E = hf$) required for that particular oscillation as Planck suggested, eventually a frequency is reached where *all* the original energy E_0 is used up at once in one *photon* of frequency radiation. In other words, some maximum frequency $f_{\max}$ will be reached with an energy which, by Planck's equation, consumes *all* of the original energy E_0 that was supplied in the experiment, or

$$E_0 = hf_{\max}$$

Since we still believe in conservation of energy, we know that we cannot extract an energy that is greater than the original energy E_0 which we put into the cavity. So the very high frequencies simply cannot be present.

The **photon** or quantum of electromagnetic energy is the basis for the success of Planck's interpretation. At $f_{\max}$ at most only one photon of that frequency could exist. At a lower frequency $f_{\max}/2$, for instance, the energy of each photon would be $\frac{1}{2}E_0$ and there could be at most two photons of that frequency. At even lower frequencies, there could be many, many quanta or *photons* of radiation, since the energy of each photon would be much smaller than the original energy E_0. Thus, according to Planck, light and all electromagnetic radiation are not emitted as continuous waves, but in bundles of energy, called photons, according to $E = hf$.

17-3

Einstein and the Photoelectric Effect

A quite different experiment, first performed by H. Hertz in the 1880's, showed that ultraviolet light could discharge a *negatively*-charged electroscope while white light could not. Furthermore, neither ordinary white light nor ultraviolet light could discharge a *positively*-charged electroscope. Albert Einstein explained this phenomenon, called the **photoelectric effect**, in another of his four papers in 1905. Beginning with Planck's hypothesis that the energy of emitted light is proportional to its frequency $E = hf$, Einstein reasoned that this peculiar experimental result could be understood if the same equation also held true for the **absorption** of light. The negatively-charged electroscope has an excess of electrons and the experiment shows that the ultraviolet light is removing electrons from the electroscope. It takes a certain minimum amount of energy W, called the **work function**, to remove a given electron from a metal. (The value of W depends upon the physical properties of the particular metal.) If the energy given to each electron is dependent upon the frequency rather than the intensity of the radiation, the phenomenon can be understood.

According to Planck's equation $E = hf$, high-frequency ultraviolet radiation gives up a large amount of energy to an individual electron in the metal. The electron is then released from the metal *if* the radiant energy $E = hf$ is *greater* than W. Visible white light is a mixture of frequencies, each lower than ultraviolet. The energy $E = hf$ given to an electron is therefore less than W and no electrons are emitted. We are assuming in this explanation that

PLANCK

Max Planck lived an orderly and precisely organized life—he played the piano for thirty minutes every day and went for walks at appointed times. He was also responsible for ideas that were to change the nature of physics markedly. □ Planck was born in 1858 in Kiel, Germany, but his family soon moved to Munich. Subsequently, he went to Berlin, where he studied under von Helmholtz and Kirchhoff, two famous physicists of the day. Traveling served to widen his view of the world, but Planck found the lecture styles of his teachers dull and did a lot of independent reading and studying. When his doctoral dissertation was largely ignored, Planck, feeling his work deserved some kind of recognition, submitted a paper on the subject to the University of Göttingen and was awarded a prize. □ Planck was very sure of his work, although he himself did not understand all its implications and was irritated that others took so long to accept it. He later said of the whole controversy that he had learned one thing: "A new scientific truth does not triumph by convincing its opponents and making them see the light, but rather because its opponents eventually die. . . ." □ Planck was an excellent musician. While he and Einstein were both in Germany, they played violin-piano duets. Einstein and Planck also shared a common enemy—Adolf Hitler. Planck, although Aryan in features and very patriotic, enraged Hitler and was dismissed from an official visit when he suggested that the Jews were being mistreated. One of his sons was later executed for conspiring against the Fuhrer, leaving Planck torn between patriotism for his country and humane feelings for his persecuted and suffering countrymen. □ Planck's sense of order also extended to his view of the world. His great respect for reason, and for the fact that through logic man could gain insight into the world, led him to devote the last years of his life to the philosophical problems of causality. □ Planck died in Göttingen in 1947 at the age of 90. He had begun working in physics at a time when it was believed that only a few loose ends needed to be tied up and the science would be completely understood. Before his death, he had witnessed an entire revolution in the way we look at the world—a revolution partially brought about by his tenacity and by his genius.

Ullstein Bilderdienst

Max Planck

17-1 Low-frequency radiation does not remove electrons; high-frequency radiation does.

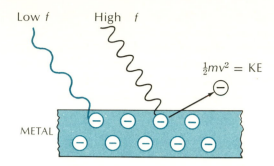

the bundles or photons of electromagnetic energy $E = hf$ are distinct and that the electron cannot "save up" the energy from several such photons. This simple picture of Einstein's completely explained the photoelectric experiments. He wrote a photoelectric equation to describe the effect:

$$E = hf = W + KE \tag{17-2}$$

where $E = hf$ is the energy of a light photon or quantum of frequency f and W is the work function of the electron in the metal. If hf is smaller than W, then *no* electron receives sufficient energy to leave the metal. If hf is greater than W, then electrons will be ejected from the metal and any excess energy $(hf - W)$ will be given to the ejected electrons as kinetic energy KE (see Figure 17-1).

Example 17-1:

The work function W of a certain metal is 3.5×10^{-12} ergs. Will violet light ($\lambda = 4.0 \times 10^{-5}$ cm) eject electrons from this metal? Will red light ($\lambda = 6.6 \times 10^{-5}$ cm)? Use $E = hf$ with $f = c/\lambda$. Then $E = hc/\lambda$, with $h = 6.6 \times 10^{-27}$ erg-sec, $c = 3 \times 10^{10}$ cm/sec, and λ (violet) $= 4.0 \times 10^{-5}$ cm. The energy of the quantum of violet light is therefore

$$E = \frac{(6.6 \times 10^{-27} \text{ erg-sec}) \times (3 \times 10^{10} \text{ cm/sec})}{4.0 \times 10^{-5} \text{ cm}} .$$
$$= 4.95 \times 10^{-12} \text{ ergs}$$

Since this value is greater than the work function of 3.5×10^{-12} ergs, **the violet light will eject electrons**.

For red light,

$$E = \frac{(6.6 \times 10^{-27} \text{ erg-sec}) \times (3 \times 10^{10} \text{ cm/sec})}{6.6 \times 10^{-5} \text{ cm}}$$
$$= 3.0 \times 10^{-12} \text{ ergs}$$

Since this energy is less than the work function of the metal, **no electrons will be emitted by the red light**.

The consequence of Planck's theory is that light or electromagnetic radiation is *emitted* from a source in the form of discrete bundles of energy called photons for which the energy is $E = hf$. The result of Einstein's photoelectric theory is that light or electromagnetic radiation is *absorbed* in materials as discrete bundles of energy or photons with energy $E = hf$. Since we can only observe the physical effects of electromagnetic energy by emitting it and/or by absorbing it, we conclude that it makes no sense to discuss electromagnetic radiation without using these quantum concepts. So we have come full circle back to Newton's particle or quantum description of light. Light and all electromagnetic radiation is not continuous in nature, but is *discrete* or *particulate*. The particle of light—the quantum of a bundle of electromagnetic energy—is the photon γ. Thus the photon (the quantum of electromagnetic energy) and the electron (the quantum of electric charge) are the first two in what will soon become a long list of the **elementary particles** of nature.

The validity of the various experiments that showed light to be a wave (Young's double-slit experiments, the Michelson-Morley experiments, etc.) is now questionable. Remember that in these experiments the interference of two light rays is understandable only if both rays are considered to be waves that could interfere with one another to produce typically observable interference patterns. But, how can light be a wave and a particle at the same time? This question has some less than trivial philosophical implications which are vaguely reminiscent of the question Galileo and others before him asked ("How can a terrestrial body move if there is no forced motion or mover to keep it in motion?") and the problem Kepler wrestled with ("How can a planetary orbit be an ellipse with the sun at one focus and nothing at the other?"). The reason these questions are pertinent to the wave-particle "duality," as it is sometimes called today, is that Galileo and Kepler were operating under the overpowering influences (the "brainwashing," if you like) of the Platonic-Aristotelian tradition. Thus, their questions were significant and, thereby, difficult to answer within the context of the accepted thinking of their time. However, they were able to provide answers to their respective questions by reaching out into the unknown far beyond Aristotelian teaching.

Likewise, schooled as we are today in the Newtonian world of mechanistic physics, we question the wave-particle duality because it is difficult to couple our mechanistic picture of a particle (a localizable portion of matter) with our ingrained mechanistic picture of a wave (a water wave, a wave on a string). Whatever light is, it sometimes behaves like what our sense experiences recognize as a wave and at other times like what they recognize as a particle. But it is probable that this duality does not actually exist in nature; more than likely it arises due to *our inability to conceptualize nature,* as when the ether "vanished" due to the null result of the Michelson-Morley experiment. Any problem associated with the wave-particle duality it would seem is ours, not nature's. We know that if the wave picture is combined with Planck's quantum equation $E = hf$, the resulting theory satisfactorily describes a photon of electromagnetic radiation.

17-4

Atoms—The Second Basis of Quantum Mechanics

When we look at a mercury vapor lamp through a diffraction grating or prism, instead of the **continuous spectrum** of colors red through violet obtained from a tungsten lamp, only certain "lines" are visible. These color lines represent **discrete** or particulate frequencies f of light emitted from the *atoms* of mercury. Our quantum picture of light then tells us that each frequency—each color line—represents an emitted photon of a particular energy $E = hf$. Now why would the mercury lamp only emit radiant energy of these particular values? It may be obvious already that this *discreteness* in the energy emitted by the mercury atoms is a new clue telling us something about atomic structure.

What are atoms? To answer this question, we turn to one of the major revolutions or breakthroughs in chemistry—the atomic theory which originated in Greek times with Democritus. The modern version began in the early 1800's with Thomas Dalton who identified chemical elements by their chemical reactions and categorized them in terms of their masses (the mass of one mole of the element, which is proportional to the mass of each constituent atom). Hydrogen was found to be the lightest atom, then helium, lithium, and so on up to uranium, the heaviest know natural element. Still later in the 1800's, Mendeleev organized these elements according to their similar chemical reactions and their weights, and arrived at what we now know as the Periodic Table of the Elements (shown in Figure 17-2). From this empirical classification scheme, he and others after him were able to predict the existence and the chemical properties of elements which had not yet been discovered. These appeared as missing entries in the periodic classification scheme.

The atomic theory was well established by 1896 when the electron was discovered to be the "particle" of negative electricity common to all atoms of all elements. Obviously, then, the electron played some important role in the structure of an atom and, possibly, could also be connected with the quantum effects observable in the discrete color lines from the mercury vapor light. The connection was not clear, however, until the structure of the

Periodic Table of the Elements*

1 H 1.0080																	2 He 4.0026
3 Li 6.941	4 Be 9.0122											5 B 10.81	6 C 12.011	7 N 14.0067	8 O 15.9994	9 F 18.9984	10 Ne 20.179
11 Na 22.9898	12 Mg 24.305											13 Al 26.9815	14 Si 28.086	15 P 30.9738	16 S 32.06	17 Cl 35.453	18 Ar 39.948
19 K 39.102	20 Ca 40.08	21 Sc 44.956	22 Ti 47.90	23 V 50.941	24 Cr 51.996	25 Mn 54.9380	26 Fe 55.847	27 Co 58.9332	28 Ni 58.71	29 Cu 63.54	30 Zn 65.37	31 Ga 69.72	32 Ge 72.59	33 As 74.9216	34 Se 78.96	35 Br 79.909	36 Kr 83.80
37 Rb 85.467	38 Sr 87.62	39 Y 88.906	40 Zr 91.22	41 Nb 92.906	42 Mo 95.94	43 Tc (99)	44 Ru 101.07	45 Rh 102.906	46 Pd 106.4	47 Ag 107.870	48 Cd 112.40	49 In 114.82	50 Sn 118.69	51 Sb 121.75	52 Te 127.60	53 I 126.9045	54 Xe 131.30
55 Cs 132.906	56 Ba 137.34	57 La 138.906	72 Hf 178.49	73 Ta 180.948	74 W 183.85	75 Re 186.2	76 Os 190.2	77 Ir 192.2	78 Pt 195.09	79 Au 196.967	80 Hg 200.59	81 Tl 204.37	82 Pb 207.2	83 Bi 208.981	84 Po (210)	85 At (210)	86 Rn (222)
87 Fr (223)	88 Ra (226)	89 Ac (227)	104 Rf (261)	105 Ha (262)													

atomic number — 1
symbol of element — H
atomic mass — 1.0080

Lanthanide Series

58 Ce 140.12	59 Pr 140.908	60 Nd 144.24	61 Pm (147)	62 Sm 150.4	63 Eu 151.96	64 Gd 157.25	65 Tb 158.925	66 Dy 162.50	67 Ho 164.930	68 Er 167.26	69 Tm 168.934	70 Yb 173.04	71 Lu 174.97

Actinide Series

90 Th (232)	91 Pa (231)	92 U (238)	93 Np (237)	94 Pu (242)	95 Am (243)	96 Cm (248)	97 Bk (249)	98 Cf (249)	99 Es (254)	100 Fm (257)	101 Md (258)	102 No (259)	103 Lr (260)

*Atomic weights of stable elements are those adopted in 1969 by the International Union of Pure and Applied Chemistry. For those elements having no stable isotope, the mass number of the "most stable" well-investigated isotope is given in parentheses.

positive electricity component—the positive ion—of the atom was determined.

Normally, the atom of any element is electrically neutral. Since it has been experimentally established that negative electrons can be removed from neutral atoms with the addition of some energy, there must be positive charges in the atoms as well. The "plum-pudding" model of the atom was widely accepted in the early 1900's. This model (see Figure 17-3) supposed that the atom consisted of a uniformly-distributed sphere of positive charge in

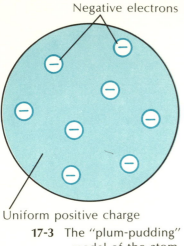

Negative electrons

Uniform positive charge

17-3 The "plum-pudding" model of the atom.

which negative electrons were imbedded like raisins in a plum pudding. Ernest Rutherford, a New Zealand physicist who had settled in England, tested this model by bombarding gold atoms (in the form of a thin gold foil) with positively-charged projectiles called **alpha particles** (the doubly-charged ions of helium obtainable from a radioactive decay). If the plum-pudding model was correct, then the helium ions would be deflected somewhat by Coulomb's law as they passed through the uniform positive charge of the atoms of gold. To observe the ions coming from the thin foil, Rutherford allowed them to hit a fluorescent screen and then watched the positions of the tiny fluorescent flashes (Figure 17-4).

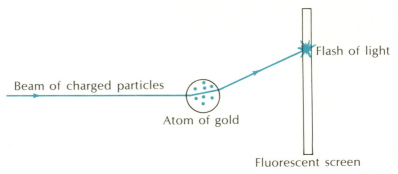

Flash of light

Beam of charged particles

Atom of gold

Fluorescent screen

17-4 Rutherford's experiment.

He found that, in addition to expected small deflections of the ions, in some cases the alpha particles were deflected at much larger angles than anticipated, and a substantial number of ions were even deflected by 180° back in the opposite direction by the uniform positive charge of the gold atoms! Rutherford calculated that the plum-pudding model, given Coulomb's law for electrostatic forces, could not have produced these results. Either he had discovered a new force law or all of the positive charge in the atoms were concentrated in one very small place. He and other physicists of the day accepted the latter explanation. Thus, in Rutherford's model, the atom was composed mostly of empty space, with the positive charges all bunched together at the center into what we now call the **nucleus** and the negative electrons spaced relatively far from the grouping of positive charges.

Since the negative electrons would normally be attracted to the positive nucleus and cause the atom to collapse, it appeared obvious that these electrons must orbit the nucleus with a centripetal acceleration provided by the Coulomb force (see Figure 17-5), in much the same way the earth orbits the sun with the centripetal acceleration provided by the gravitational force (see Chapter 6, pages 80–81).

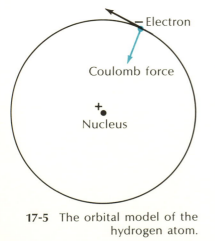

Electron

Coulomb force

Nucleus

17-5 The orbital model of the hydrogen atom.

In order to study Rutherford's model quantitatively, we choose the atom of the lightest element—hydrogen—in the hope that it will be the simplest. The experimental information about hydrogen available in Rutherford's time was that its atom contained only one electron and, therefore, its central nucleus contained only one positive charge (today called the **proton**). The **line spectra** of the light emitted by hydrogen had already been observed in the 1800's, and an empirical formula (like Kepler's laws were empirical formulas) was devised by trial and error to fit the hydrogen lines by a Swiss high school teacher, Johannes Balmer, in 1885. Balmer's formula was

$$f = C\left(\frac{1}{2^2} - \frac{1}{n^2}\right) \quad , \quad n = 3,4,5\ldots \tag{17-3}$$

where f is the frequency of the observed color in the hydrogen spectrum, C is a constant, and n is any integer greater than 2 (i.e., $n = 3,4,5\ldots$). The first line in the hydrogen spectrum is red and its frequency is obtained by letting $n = 3$. The frequency of the second line, a green color, is obtained with $n = 4$; the third one, violet, with $n = 5$; and so on. Agreement between the observed frequencies and those predicted by the Balmer formula was astounding. The subsequent Bohr–de Broglie theory of the hydrogen atom showed why it worked so well.

Niels Bohr made the next enormous break with traditional classical physics in a way reminiscent of Copernicus' break with Aristotelian physics. Copernicus postulated new rules for the operation of the celestial spheres by assuming a sun-centered universe, although he still used the perfect circles of Aristotelian astronomy. Bohr postulated new rules for the operation of elementary particles (the electron and proton in the hydrogen atom) although he still retained many of the concepts of Newtonian mechanics. In 1913, Bohr's model for the hydrogen atom was published; it provided what we now think of as a halfway theory, a semimechanistic picture of the atom. Bohr states in the beginning of his paper:

In order to explain the results of experiments on scattering of α rays by matter, Prof. Rutherford has given a theory of the structure of atoms. According to this theory, the atoms consist of a positively-charged nucleus surrounded by a system of electrons kept together by attractive forces from the nucleus.

This paper is an attempt to show that Rutherford's atom-model affords a basis for a theory of the constitution of atoms. It will further be shown that from this theory we are led to a theory of the constitution of molecules. In the present first part of the paper, the mechanism of the binding of electrons by a positive nucleus is discussed

17-5

The Hydrogen Atom and Niels Bohr

in relation to Planck's theory. It will be shown that it is possible, from the point of view taken, to account in a simple way for the [Balmer] law of the line spectrum of hydrogen.[1]

The startling success of Bohr's work was that it showed how Balmer's empirical formula (Equation (17-3)) could be derived from the more fundamental laws of electrostatic attraction by Coulomb's law and centripetal force. In his derivation, Bohr set forth a postulate so new that it was hard for many to believe at that time—that the **angular momentum** (mvr) of the orbiting electron could have only **discrete** values (i.e., any *integer multiple* of Planck's constant h divided by 2π). In mathematical terms, Bohr postulated the new rule for playing the atoms game:

$$\text{angular momentum of electron} = mvr = n\left(\frac{h}{2\pi}\right) \qquad \textbf{(17-4)}$$

where n is an integer ($n = 1, 2, 3, 4 \ldots$), m is the mass of the electron, v is the velocity of the electron, and r is the radius of the electron orbit from the proton nucleus. What Bohr proposed was that *angular momentum is quantized;* that is, the angular momentum

[1] Niels Bohr, *Philosophical Magazine*, **26** (1913).

The Niels Bohr Library, American Institute of Physics

*Niels Bohr
with his five sons.*

BOHR

Niels Bohr loved people. His entire life centered around his deep involvement with other human beings, beginning with his family and ending in an amazing place—an institute devoted to thought. ▫ Bohr was born in Copenhagen in 1885, and there he grew up in a close, creative atmosphere with his brother Harold. Later, when both were members of a discussion group on philosophical problems, the talks frequently became dialogues conducted solely by the Bohr brothers. Their thinking was very similar and they clarified each other's ideas. ▫ Bohr became excited about science very early in his life. One day the high school laboratory was virtually rocked by a series of explosions. An instructor said "It's Bohr."; he was right. ▫ After an education marked by brilliance, Bohr went to Cambridge to study under J. J. Thomson. (Bohr introduced himself to the great physicist by pointing out a few errors in Thomson's work.) Bohr's stay at the Cavendish Laboratory in Cambridge was uneventful, except for the fact that there he met Ernest Rutherford who had just formulated a new theory of the atom. Bohr was fascinated with Rutherford and Rutherford was favorably impressed with Bohr, so, in the spring of 1911, Bohr went to work with Rutherford at Manchester. ▫ There Bohr combined Rutherford's discovery of the atomic nucleus with the quantum concepts of Planck and Einstein. The resultant Bohr theory of the atom astounded the world of physics in 1913 and Bohr was awarded the Nobel Prize in 1922 for his work. ▫ In 1916, Bohr was offered the Professorship of Theoretical Physics in Copenhagen. There, after a long struggle, Bohr opened the Institute for Theoretical Physics in 1921; it soon became a center of intellectual activity for physicists from all over the world. The Institute was completely informal, mixing ping-pong and relativity, Faust (Bohr loved *Goethe*) and quantum mechanics, western movies and electron theory, in a relaxed atmosphere of creative thinking unparalleled anywhere else in the world. Bohr guided the whole project with love, compassion, and kindness. ▫ Also highly concerned with politics, the Institute's strongest period came in the late thirties. When the Nazis invaded Denmark in 1940, Bohr decided to remain in Copenhagen as long as was possible; he finally escaped just after British intelligence discovered he was to be deported to Germany. ▫ Bohr first went to Sweden, where he tried to convince the government to help the Danish Jews, and finally came to the United States and began work on the atomic bomb. He devoted the years after World War II to attempting to make the nations of the world see the insanity of continuing the situation which had precipitated it. He worked with the United Nations and devoted his time to the peaceful application of atomic energy. During his last years, his keen mind turned toward the subject of biology. ▫ Bohr died in 1962. His ideas in physics had already changed the world in many ways, but his concept of international cooperation was unfortunately still a long way from being effected.

of the orbiting electron cannot have any arbitrary value (as do the planets orbiting the sun), but only certain well-defined values: $(h/2\pi)$, $2(h/2\pi)$, $3(h/2\pi)$, $4(h/2\pi)$, and so on. Bohr explains

> If we therefore assume that the orbit of the electron is circular, the result of the calculation can be expressed by the simple condition: that the angular momentum of the electron round the nucleus is equal to an entire multiple of a universal value, independent of the charge on the nucleus.[1]

DE BROGLIE   Louis de Broglie was born in Dieppe, France, in 1892. After receiving his Baccalaureate in 1909, his interests inclined toward literary studies, and he took his licenciate in history in 1910 at the Sorbonne. Soon after, however, he became interested in science and began studying for the licenciate in science which he obtained in 1913.   □   When World War I broke out, de Broglie was attached to a radio-telegraphy unit where he remained from 1914–1918. At the end of the war, de Broglie resumed his studies both in general physics and in the experimental work pursued by his older brother Maurice. Louis was particularly interested in theoretical problems involving quanta which had already been so successfully used by Bohr ten years earlier. In 1924, de Broglie presented his thesis entitled "Research in the Quantum Theory" to the University of Paris. The novelty of his ideas—the result of about two years' previous work—astounded the world of physics. But de Broglie's work was quickly accepted. Only one year later, it served as the basis for the development of the general theory known as "wave mechanics," which was to completely transform the knowledge of the physical phenomena of the atomic scale.   □   de Broglie continued to publish original works on the new wave mechanics in addition to teaching broad courses at the Sorbonne. After two years, he began teaching courses in theoretical physics at the Institut Henri Poincaré which had just been opened in Paris for the purpose of insuring the teaching and development of mathematical and theoretical physics. Holder of the Chair of Theoretical Physics in the Faculté des Sciences of Paris since 1932, de Broglie has subsequently given a course on a *different* subject at the Institut Henri Poincaré every year—a difficult accomplishment almost unheard of in American universities. de Broglie retired from the Faculté des Sciences in 1962.   □   Since 1951 de Broglie resumed the study of his former attempt in 1927 to give wave mechanics a causal interpretation in the classical framework of space and time, an attempt which he had abandoned earlier due to the general acceptance by most physicists of the day of the probabilistic interpretation of Born, Bohr, and Heisenberg. de Broglie's renewed work has led to a number of new and encouraging results which he continues to publish.   □   de Broglie has received many awards for his work in physics research and for his work in teaching—most notably the Nobel Prize awarded in 1929 and the 1952 UNESCO Kalinga prize for his work in scientific popularization. At present, he is continuing to research interpretational problems in quantum mechanics at his home in Neuilly-sur-Seine, France.

17-6

de Broglie and Electron Waves

Bohr's theory was accepted by many physicists, partly because of the successful derivation of the Balmer formula for the hydrogen spectrum; many others ignored his theory until 1924. In that year, a young graduate student, Louis de Broglie, submitted his PhD thesis to the University of Paris. The thesis proposed a fundamental reason for Bohr's postulate, Equation (17-4), and provided the one last clue that led to the final formulation of quantum mechanics in 1925–26.

de Broglie drew an analogy between electron waves and light photons which sometimes behave like particles and sometimes behave like waves. He suggested that the electron—a material particle—was itself a wave! If we can accept that a wave (electromagnetic radiation) can be a particle (photon), then it should not be too difficult to accept the wavelike nature of what we once assumed to be material particles. In 1927 an experiment similar to Young's double-slit experiment for light was performed for electrons; it showed clearly that electrons, the particles of electricity, **interfere** with one another. From our past sense experience, we know that only *waves* can produce the in-phase and out-of-phase relations that yield interference patterns, so we must conclude, with de Broglie, that the electron is a wave.

What then is the wavelength of this particle-wave? de Broglie used the analogy of a light wave-photon to determine it. A light photon has an energy $E = hf$, where f is the frequency and h is Planck's constant. If we substitute the relation between wavelength λ and frequency f ($f\lambda = c$), then Planck's equation in terms of the wavelength is $E = hc/\lambda$. But, according to Einstein's relativity, anything that has energy also has mass ($E = mc^2$). So de Broglie equated these two energies, one from the quantum of light and the other from relativity:

$$E = mc^2 = \frac{hc}{\lambda}$$

or

$$mc = \frac{h}{\lambda} \tag{17-5}$$

If we then solve for λ, we find

$$\lambda = \frac{h}{mc} \tag{17-6}$$

which would normally give us the wavelength λ of a light photon in terms of Planck's constant and the "mass" of that light photon. de Broglie recognized the product of the mass m of a light photon times its velocity c as the momentum p of the photon; therefore he set $\lambda = h/\text{momentum} = h/p$. He made a great intuitive leap when he then *associated this relation between wavelength λ and momentum p with electrons.* He wrote

$$\lambda = \frac{h}{p} = \frac{h}{mv} \tag{17-7}$$

for the wavelength of an electron or any "material" particle of mass m traveling at velocity v. Reportedly, no one at the University of Paris could understand de Broglie's thesis, so it was given to Einstein to read. Einstein also admitted that he could not understand it, but suggested it might be very important and that de Broglie's PhD should be awarded. It was.

The consequences of de Broglie's hypothesis were far-reaching,

Académie des Sciences, Paris

Louis de Broglie

both to the practitioners of the Newtonian-Maxwellian classical physics and to those of the mechanistic philosophy and world picture that had held sway for nearly 300 years after Newton.

Several consequences resulted from de Broglie's construct or theory of Equation (17-7). First, it "explained" Bohr's postulate on the quantization of angular momentum, as we will see in the next section. Secondly, it provided a basis for performing quantitative calculations about the internal structure, the radiation emission, etc., of atoms and nuclei. (If particles are waves, they must obey or satisfy a wave equation which can be used quantitatively in calculations, like the equation for electromagnetic waves.) Thirdly, if particles are waves and obey a wave equation, then they are no longer localizable at a particular point in space as previously thought (the uncertainty principle discussed in Chapter 18 follows from this). With de Broglie's theory, the deterministic philosophy of the nineteenth century, built on the foundation of a Newtonian mechanistic world view, had nearly collapsed. When the contributions of Einstein's relativity and the Planck-Einstein quantum theory of light are added, we begin to grasp the enormity of the changes that physics went through in the first quarter of the twentieth century. The very foundations and roots of everything physicists had previously believed crumbled.

17-7

de Broglie Explains Bohr

If an electron is a wave, then we can no longer picture it in the hydrogen atom as a localizable particle traveling around the proton in an orbit. The picture must be changed to one in which the electron can exhibit its wave properties. The significant wave property missing in our old picture is that a wave can interfere with itself or with other waves. A wave on a rope which is continuously shaken travels down the rope to the end, reflects back to the starting point, reflects again to travel down the rope, and so on. The only waves that can permanently exist on the rope are those which have a shape down the rope identical to their shape back up the rope. These are called **standing waves**; as they travel down and up and down the rope again, they will always retrace themselves. If the length of the rope is L, then such a retracing is possible for wavelength: $\lambda = 2L$ (see Figure 17-6 a) (this situation is similar to the vibration of a piano string). The retracing is also possible for a wavelength $\lambda = L$ (Figure 17-6 b) and, in general, for those values of the wavelength $\lambda = 2L/n$ (Figure 17-6 c), where n is an integer ($n = 1,2,3 \ldots$). Any other wavelengths besides these "quantized values," after many retracings, will cancel themselves out by destructive interference.

de Broglie's concept of the electron in the hydrogen atom is shown in Figure 17-7. The "orbit" of the electron has a radius r.

The condition that permits the electron wave to retrace itself over and over again is that the length of the orbit (i.e., the circumference of the orbit $2\pi r$) must be equal to a whole number of wavelengths λ. Thus $2\pi r = \lambda$, 2λ, 3λ, 4λ, etc., or $2\pi r = n\lambda$ where n is an integer. If we stretch out the orbit shown in Figure 17-7 so that it all lies on a straight line, we can draw the possible electron wavelengths that will retrace themselves according to $2\pi r = n\lambda$ (Figure 17-8). With the result $2\pi r = n\lambda$ for the possible electron orbits that do not cancel themselves out along the circumference of the atom but retrace themselves orbit after orbit, de Broglie added Equation (17-7) $\lambda = h/mv$ to $2\pi r = n\lambda$:

$$2\pi r = \frac{nh}{mv} \tag{17-8}$$

If we multiply both sides of this equation by mv and divide by 2π, we derive Bohr's postulate [Equation (17-4)], that the angular momentum is quantized in multiple units of $(h/2\pi)$:

$$mvr = \frac{nh}{2\pi}$$

Beginning with Equation (17-4), only a small amount of algebra is required to derive the Balmer formula for the frequencies of the line spectra of hydrogen. Briefly, the classical total energy (kinetic plus potential) of the electron is calculated as a point particle traveling around the proton in a circular orbit. The expression obtained for the energy E_n of the electron in a quantum orbit denoted by the integer n is

$$E_n = -\frac{me^4}{2\hbar^2} \times \frac{1}{n^2} \tag{17-9}$$

In this equation, m is the mass of the electron (9.1×10^{-27} g), e is the charge on the electron (4.8×10^{-10} stat-C), and $\hbar$ is the symbol for $(h/2\pi)$ and has the value $h = 1.05 \times 10^{-27}$ erg-sec. The energy E is in ergs. Note that the total energy in the electron in the hydrogen atom is *negative*. The reason for this is that we define the total (nonrelativistic) energy of the electron and proton when they are infinitely far apart as equal to zero. When the electron is attracted to and is in a stable orbit around the proton, its energy is less than its zero energy at infinity. Thus, the total energy (KE + PE) is negative by definition. The electron is bound to the proton or, in quantum language, the electron is *in a* **bound state** as denoted by the **quantum number n**.

The spectrum of the discrete lines of light emitted by hydrogen can be obtained directly from Equation (17-9), which gives the values for what are called the **energy levels** of hydrogen. The electron is normally in its **ground state** with an energy

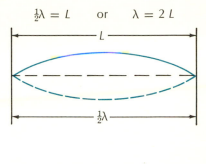

$$\tfrac{1}{2}\lambda = L \qquad \text{or} \qquad \lambda = 2L$$

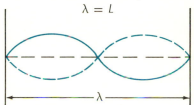

$$\lambda = L$$

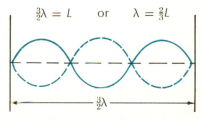

$$\tfrac{3}{2}\lambda = L \qquad \text{or} \qquad \lambda = \tfrac{2}{3}L$$

17-6 Allowed wavelengths in a string fixed at both ends.

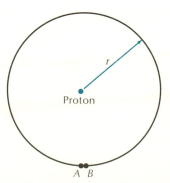

17-7 de Broglie's orbit of the electron in the hydrogen atom. The length from A to B is $2\pi r$.

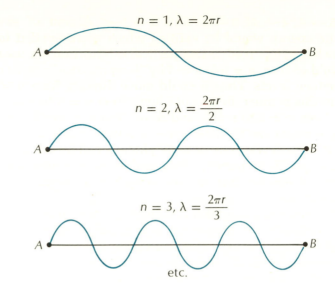

$n = 1, \lambda = 2\pi r$

$n = 2, \lambda = \dfrac{2\pi r}{2}$

$n = 3, \lambda = \dfrac{2\pi r}{3}$

etc.

17-8 Possible electron waves along the orbit of the circumference of the hydrogen atom.

$E_1 = -me^4/2\hbar^2$, where $n = 1$. This is the lowest possible energy the hydrogen electron can have. If we add electrical energy or heat (which is energy of motion) to the hydrogen atom, we can excite the electron from the ground state $n = 1$ to a higher, **excited state** denoted by any of the quantum numbers greater than one ($n = 2, 3, 4, \ldots$). Assume that the electron is excited to an energy level corresponding to $n = 2$. Then the energy in this level is

$$E_2 = -\frac{me^4}{2\hbar^2} \times \frac{1}{2^2} = \frac{1}{4} E_1$$

or greater (less negative) than the energy E_1 of the ground state. Since the electron will normally attempt to return to the ground state or to the lowest possible energy, it will move naturally from the $n = 2$ level to the $n = 1$ level. To do so, it must get rid of the excess energy that the $n = 2$ level has over the $n = 1$ level or an energy of $E_2 - E_1$. The electron emits a single photon, a quantum of light, whose energy $E = hf$ is exactly equal to the energy difference between energy levels $n = 1$ and $n = 2$, $E_2 - E_1$ or

$$E = hf = E_1 \left(\frac{1}{1^2} - \frac{1}{2^2} \right) = \frac{3}{4} E_1 \tag{17-10}$$

In general, the electron in the original ground state atom, with sufficient energy, could have moved up to an $n = 3$, $n = 4$, $n = 5$, or an even higher n-level. For example, if it reached an $n = 5$ energy level, its energy would be $E_5 = \frac{1}{25} E_1$ and it could return to the ground state $n = 1$ in a variety of ways. It could return to $n = 4$ first, then to $n = 3$, $n = 2$, and $n = 1$. In *each* transition—$5 \rightarrow 4$, $4 \rightarrow 3$, $3 \rightarrow 2$, and $2 \rightarrow 1$—a photon of electromagnetic energy would be emitted which corresponded to the energy change involved in that particular transition. Or the electron could go from $n = 5$ to $n = 1$ by "jumping over" one or more intermediate levels. Thus it could move directly from $n = 5$ to $n = 1$ in this manner, or it could jump from $n = 5$ to $n = 2$ and thence from $n = 2$ to $n = 1$, and so on. You can see that a large variety of ways exist for the electron to return from $n = 5$ to $n = 1$ and that, in each way, one or more photons will be emitted, each having a specific frequency. This enables us to understand the existence of discrete line spectra in hydrogen, mercury, and in all other atoms.

Bohr knew his proposal that a single photon be emitted from the excited levels of hydrogen was new and went against the usual ideas of classical theory. "The assumption is in obvious contrast to the ordinary ideas of electrodynamics, but appears to be necessary in order to account for experimental facts," he said, thereby breaking completely with the traditional theories of classical physics.

To picture the energy changes just described, we will draw an "energy-level diagram." Energies of the stable levels of hydrogen are plotted vertically, beginning with zero energy at the top. The length of the arrow (shown in Figure 17-9 for an $n = 3$ to $n = 1$ transition) gives the relative amount of energy (or frequency) of the electromagnetic radiation emitted. The particular series of color lines that Balmer found for hydrogen happen to be those transitions from higher levels that end on $n = 2$. These transitions—$3 \rightarrow 2$, $4 \rightarrow 2$, $5 \rightarrow 2$, and so on—also just happen to have the right energy so that the frequencies of the emitted light quanta,

17-8

Energy-Level Diagrams

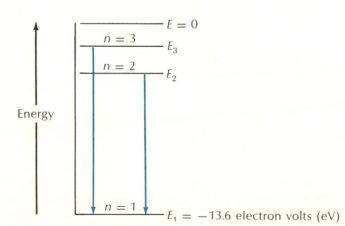

17-9 Energy levels of the hydrogen atom.

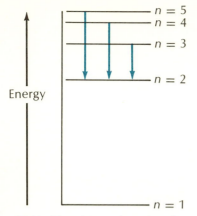

$n = 5$
$n = 4$

$n = 3$

$n = 2$

Energy

$n = 1$

17-10 Transitions that give the Balmer Series of color lines for hydrogen.

the photons, are in the *visible* region of the electromagnetic spectrum (see Figure 17-10).

The Bohr–de Broglie formula for the energy levels, Equation (17-9), would predict the energy of the photon to be $E_n - E_2$, where $n = 3, 4, 5, \ldots$.

$$\text{photon energy} = (E_n - E_2) = hf$$

and, with the help of Equation (17-9), the frequencies of the observed color spectra in hydrogen are

$$f = \frac{me^4}{4\pi\hbar^3}\left(\frac{1}{2^2} - \frac{1}{n^2}\right) \tag{17-11}$$

which is exactly the same as Balmer's empirical formula, Equation (17-3). (The constant C in Balmer's formula is now seen to be derivable from the fundamental quantities m, e, and $\hbar$.) In addition to the series of line spectra given by the Balmer formula, we can see that another entire series of lines is obtained when hydrogen atoms make transitions from higher levels to $n = 1$. The energies of these photons are much greater than those of the Balmer series; consequently, the frequencies are higher and the radiation lies in the ultraviolet region outside the visible spectrum. (This has been confirmed by ultraviolet photography techniques.) An entire series of lines due to transitions from higher levels to $n = 3$ also exists. Their frequencies are lower than visible light and lie in the infra-red spectrum. (They have also been observed to agree with theoretical predictions.)

We have seen that the de Broglie hypothesis of particle waves leads to the Bohr postulate of the quantization of angular momentum and that the Bohr theory explains the line spectra emitted by the hydrogen atom. In the next chapter, we will see how these ideas led to the final resolution of the quantum revolution, only a year after de Broglie's hypothesis of the wave nature of particles.

Questions

1 In what sense do we think of light as waves? In what sense do we think of light as particles?

2 What is the photoelectric effect? Viewing light as particles, how did Einstein explain this effect?

3 What was the "ultraviolet catastrophe"? Explain Planck's hypothesis to eliminate this catastrophe.

4 Describe the "plum-pudding" model of the atom.

5 How did Rutherford's experiment prove that the plum-pudding model was false?

6 Niels Bohr proposed a model of the atom consistent with Rutherford's findings. Describe it.

7　What are the Balmer lines? How did Bohr's atom explain them?

8　What remarkable hypothesis did Louis de Broglie make about matter? How did this hypothesis help to explain the stability of the atom?

9　In what sense did de Broglie's hypothesis build upon previous work? In what sense was it completely new?

10　Explain how interference patterns of electrons (a diffraction pattern, for example) might be used to check de Broglie's formula for the wavelength of "particles" $\lambda = h/p$.

Questions Requiring Outside Reading

1　Explain briefly why "classical" (i.e., nonquantum mechanical) electricity and magnetism (Maxwell's equations) predict that the Bohr model of the atom is unstable.

2　What reasoning led Isaac Newton to propose a "corpuscular" theory of light? For what reasons did both Einstein and Planck propose such a theory?

3　Briefly outline the history of the particle theories of light from the ancient Greeks through Newton to Einstein. What changes led to each new theory?

4　Describe some phenomena which can be better or more simply explained by the *particle* theory of light. Explain why each is more consistent with a particle theory.

5　Repeat Question 4 for the *wave* theory of light.

Problems

1　What is the Planck-Einstein formula for the relationship between the energy of a photon and its frequency? Use this formula to find the energy of a photon with a frequency of 1.5×10^{10} sec^{-1}.

2　What is the frequency of a photon with an energy of 10^{-16} ergs?

3　Describe the relationship between the energy of a photon and its wavelength. What is the energy of a photon with a wavelength of 3×10^{-10} cm?

4　What is the wavelength of a photon with an energy of 6.67×10^{-27} ergs?

5　Green light causes a photoelectric effect with a certain metal, but yellow light does not. Will red light cause a photoelectric effect with the same metal? Will blue light?

6　What is the relationship for the *minimum* frequency of light necessary to produce a photoelectric effect from a metal with a work function of W? Use this relationship to find the minimum frequency for a metal with a work function of 3.3×10^{-12} ergs.

7　What is the longest wavelength of light which will cause a photoelectric effect from a metal with a work function of 10^{-11} ergs?

8　A hydrogen atom in the state $E_3 = -1.5$ eV makes a transition to the state $E_2 = -3.4$ eV. What is the energy of the photon which the atom must emit to achieve this transition? What is its wavelength? Will this light be visible? (The symbol eV stands for the energy unit **electron-volt**, used in atomic and nuclear physics. *Note:* One eV $= 1.6 \times 10^{-12}$ ergs.)

9 An ionized hydrogen atom captures an electron which "falls" down to the state $E_2 = -3.4$ eV. What is the wavelength of this light? What is its color?

10 What is de Broglie's formula for the wavelength of a particle? What is the wavelength of a proton traveling at 100 cm/sec? Of an electron at the same *speed*? (The mass of a proton is 1.67×10^{-24} g; of an electron, 9.11×10^{-28} g.) (*Hint:* Use $p = mv$.)

11 What is the de Broglie wavelength of a 250 g baseball traveling at 2×10^3 cm/sec?

Quantum Mechanics: A New Picture of the Universe

18

In 1925–26, two independently devised theories, one proposed by Werner Heisenberg and the other by Erwin Schrödinger, provided the "rules of the game" which were to lead to the resolution of the quantum mechanics revolution. Since the two theories were subsequently shown to be mathematically equivalent, we will examine only Schrödinger's picture of quantum mechanics—what he called "wave mechanics"—in this chapter. The de Broglie proposal that electrons and indeed all "material" particles have wavelengths $\lambda = h/mv$ [Equation (17-7)] induced Schrödinger to write a wave equation for electrons. His partial differential equation, which can be solved only with methods of the calculus, gives a detailed quantitative description of the time development—that is, the changes in motion of the particle-waves when two or more particles interact with one another. Because solving the Schrödinger equation is beyond the mathematical level of this book, it will be discussed in a qualitative way here.

18-1
Introduction

The equation that Schrödinger proposed is nonrelativistic. It is not based upon Einstein's relativistic description of space, time, mass, and energy, but upon the classical description and therefore only pertains to particles with velocities which are small in comparison with the velocity of light. For most problems in atomic and nuclear physics, Schrödinger's nonrelativistic approximation to the relativistic world is satisfactory in that it can be used to solve many, many problems. As we will see in Chapter 20, the Schrödinger equation fails only when it is used to describe the interactions of the so-called "elementary particles."

18-2
The Schrödinger Equation

SCHRÖDINGER

Erwin Schrödinger was born in 1887 and grew up in Vienna, the only child in a family with broad academic interests. His father was highly gifted and studied chemistry, botany, and Italian painting all in depth. Young Schrödinger, obviously influenced by his scholarly environment, studied the sciences and German poetry independently, although during his school years at the Gymnasium he expressed his dislike of memorizing and learning from books—a distaste he shared with many other famous physicists. □ In 1910 he received his diploma from the University of Vienna. His subsequent career in physics was interrupted when he served as an artillery officer during World War I, but after the war Schrödinger resumed his academic pursuits, teaching first at Stuttgart, then at Breslau, and finally at Zurich. During his last years in Zurich, he began his creation of wave or quantum mechanics with the now famous "Schrödinger Equation." In 1927 Schrödinger went to Berlin. Later he was awarded the Nobel Prize in 1933 (with Dirac) for his earlier work on wave mechanics. □ Schrödinger had succeeded Planck at the University of Berlin by the time of Hitler's rise to power. Deciding he could no longer remain in Germany, Schrödinger moved to England for a while and then to Graz, Austria, in 1936. But with the annexation of Austria in 1938, his arrest as an "unfriendly" citizen appeared imminent. He just managed to escape, finally settling in the Institute for Advanced Studies in Dublin, Ireland. □ Schrödinger's later interests were in biology and in the foundation of quantum physics. Always unconventional in his personal habits, he carried his belongings in a packsack on his back instead of using a suitcase. He preferred working by himself, rather than with students or colleagues. After his retirement from the Dublin Institute in 1955, Schrödinger returned to his childhood home in Vienna. He died there in 1961.

Erwin Schrödinger

The Niels Bohr Library, American Institute of Physics

Particles interact with one another other by means of known physical forces such as the gravitational force or the electrodynamic forces. Atoms and molecules, in which electrons are attracted by the positive charges on the nucleus, interact due to the force given by Coulomb's law. The Schrödinger equation is in a way the quantum equivalent of Newton's second law $\vec{F} = m(\Delta \vec{v}/\Delta t)$ which gives the dynamics of classical particle motion (that is, the changes in velocity $\vec{v}$ with time t under the influence of a known force $\vec{F}$). Likewise, Schrödinger's wave equation gives us the quantum dynamics; it tells us how the wave amplitudes of a particle change with time under the influence of a known force. This **particle-wave amplitude** ψ gives specific numerical values for the amplitude of the particle-wave that vary from point to point in space. Although in the past there have been some differences of opinion about the physical interpretation of ψ, it is now generally accepted that ψ^2, the square of a particle's wave amplitude at a given point in space, equals the **probability** of finding the particle at that particular point. This interpretation is similar in spirit to that of macroscopic waves like water waves or sound waves, where the square of the wave amplitude is proportional to the energy of the wave at that point.

ψ Psi

The de Broglie–Bohr orbital picture of the atom is thus replaced by a new picture that the solutions of the Schrödinger equation provide. This has been justified, as have all theories in physics, by the enormous success of this equation in predicting experimental results. It is still tested over and over again, every day, in many laboratories around the world and no observed disagreement has yet been found in atomic, molecular, and low energy nuclear physics applications. Bohr pictured the atom as consisting of a localizable electron orbiting the nucleus at a radius r; de Broglie changed this picture so that the particle became a wave along that orbit, still located at distance r from the nucleus. The wave amplitude ψ of the Schrödinger equation provides a different picture still: The electron is "smeared out" like a wave all along the circumference of the orbit as in de Broglie's picture, but its distance from the nucleus, rather than being a sharply defined orbit of radius r, is also "smeared out" over all distances. There is always some probability of finding the electron anywhere in space, from a region near the nucleus to distances out to infinity. However, the *probability* is greater that the electron will be "found" near the classical orbit radius r of de Broglie and Bohr than anywhere else.

Although quantum mechanics gives us a smeared-out, probabilistic picture of the location of particles in space, the energies of these particles still have discrete, well-defined values similar to those obtained from Bohr's orbital picture. Any two or more particles that exert an attractive force on one another will have a wave amplitude ψ that generally has at least one and usually many bound states or quantum energy levels at negative energy. And as before, an electron transition from one energy level to another lower level results in the emission of a photon of electromagnetic radiation with an energy hf equal to the energy difference of the two levels. Schrödinger's quantum mechanics is more successful than the semiclassical Bohr–de Broglie picture because it can predict not only the positions of the discrete spectral lines of atoms, but also the intensities or relative brightnesses of the various lines which the earlier Bohr–de Broglie model could not. The semiclassical picture was also unable to deal with problems involving collisions or scattering of particles.

Thus far, our discussion of quantum mechanics has centered around the case in which one particle, the electron, is bound to another, the nucleus. This bound state occurs because (1) the force between the two particles is *attractive* rather than repulsive, and (2) the total energy of the two bound particles is *negative* (that

18-3

Quantum Mechanical Collisions

is, less than the zero of energy the two particles would have if they were far apart). But what happens if the force between the particles is repulsive or the energy between them is greater than zero? In these cases, the two particles do not form a bound state. If the energy is greater than zero (that is, if the energy is positive), the particles have a net positive kinetic energy. In this case, the two particles collide or, in current terminology, they **scatter** from one another (a situation analogous to the classical scattering of two billiard balls).

The wave or quantum mechanical description of this collision or scattering process usually begins with a particle of mass m and velocity v *impinging* on the target from which it will eventually scatter. We begin with this description because we can relate it to the physical process in which some type of particle accelerator (electron gun or proton cyclotron) provides particles of a known mass m with a definite energy E which is related to their velocity v. So far this sounds like a classical Newtonian picture; however we also require that the particle behave like a wave of wavelength $\lambda = (h/mv)$. We do this by constructing what is called a **wave packet**—a short train of waves a few wavelengths long (see Figure 18-1)—to describe the particle. (The method by which this packet is constructed will be discussed in the next section.)

The wave packet or particle then interacts with another particle (Figure 18-2) via some force law—Coulomb's law in the case of electrons and other charged particles in an atom or the strong nuclear force in the case of **nucleons** in an atomic nucleus. The incident or striking particle collides with the target (the atom or the nucleus, for example) and is bounced off or scattered in a manner characteristic of the particular mutual force law of the two particles. Instead of the incident particle being localizable and scattering in a particular direction, quantum mechanics tells us that a wave is sent out in all directions (Figure 18-3), just like a water wave is sent out when a pebble is dropped into a pond. Unlike the water wave, which is the same in all directions, however, the particle wave will have different amplitudes in different directions. It will also be three dimensional, while the water wave travels outward in only two dimensions. The square of the ampli-

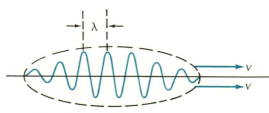

18-1 A wave packet traveling in the right with velocity v.

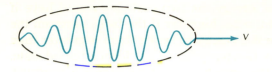

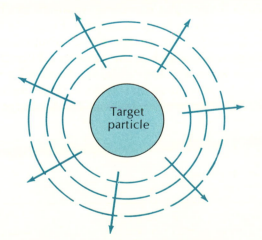

18-2 A wave packet about to hit the target particle.

tude of the particle wave in any given direction gives the probability that the "particle" will be found going in that particular direction. The wave function ψ which contains all this information is found by solving the Schrödinger equation for the particular kind of particles with their specific velocities and force laws.

When an electron collides with an atom, we may find that the wave function ψ, and hence the probability ψ^2 of finding the electron, is greater at one angle than at another. For instance, it might happen that our solution ψ of the Schrödinger equation predicts that electrons should be found with the largest probability at an angle around 30° from the original direction of the incident electrons. We then perform the experiment, with many billions of electrons colliding with many billions of target atoms (all of the same type) and, if our theory is a good one, we actually find more electrons at an angle of 30° than at any other angle.

If the experiment does not agree with our theoretical predictions, as is often the case, the force law used in the Schrödinger equation must be changed. This does not invalidate the rules of quantum mechanics; we simply did not use quite the correct force law in the equation. This occurs often because, although we know the basic forces (Coulomb's law, the nuclear force law) that can

18-3 Scattered waves are sent out in all directions.

HEISENBERG

Werner Heisenberg was born in Würzburg, Germany in 1901. His family later moved to Munich, where he was enrolled in the same school that Planck had attended. Heisenberg enjoyed his classes (notably Latin and Greek). He soon became interested in mathematics and studied differential calculus at the age of twelve. Eventually he began to take an interest in describing the world around him in mathematical terms and this encouraged him to delve more deeply into the subject. □ Heisenberg was only 17 when World War I ended, but he served in the troops fighting the revolution in postwar Germany. This gave him a certain amount of freedom, at least from parents and teachers, and he spent a lot of time on the roof of army headquarters reading Plato. After his release, he entered the University of Munich where he earned his Ph.D. □ From the university, Heisenberg went to Göttingen as an assistant to the famous Max Born and here he heard one of Niels Bohr's lectures on physics. During the question and answer session, Heisenberg told Bohr that he felt that Bohr's arguments were weak. After the session, Bohr invited Heisenberg to go for a walk to discuss his points in greater detail. Shortly afterwards, Heisenberg went to Copenhagen on a grant from the Rockefeller Foundation to work in Bohr's center for the study of atomic physics. □ A highly intuitive physicist, Heisenberg once said "... I must start not from detail but from ... a feeling I have about the way things should be." He usually published his work rather quickly, often leaving it to others to prove that his conclusions were correct. □ In 1925, on a small island in the North Sea between cliff-climbing expeditions, Heisenberg developed what was to be his Nobel prize-winning theory. When he returned to Göttingen, he handed his theory to Max Born who suggested that Heisenberg's mathematical scheme could be equivalent to matrix algebra. Heisenberg learned the necessary matrix algebra and founded the matrix mechanics which Schrödinger later showed to be equivalent to his own wave mechanics of 1926. In 1932, Heisenberg was awarded the Nobel Prize for his work on quantum mechanics. □ During World War II, Heisenberg published papers on the so-called S-matrix, an important tool for the description of elementary particle physics, and supervised Germany's attempts to construct nuclear reactors. In 1945 he was taken prisoner by the allied forces and detained in England for a while. After the war, he returned to Göttingen to reorganize the Institute for Physics there. Later, Heisenberg moved back to Munich, where he still resides, to begin work on a field theory of elementary particles.

exist between two particles, when there are many particles the problem is just too difficult mathematically to be solved in terms of these basic **two-body forces**, even with the aid of the largest computers. In fact, the quantum mechanical "three-body" problem has been solved only recently.

When there are hundreds of particles, all interacting with one another due to some well-defined, two-body force law, the problem is at present impossible to solve exactly. So we construct a model for the combined forces of all the particles and insert the force model into the Schrödinger equation. Intuition, ingenuity, and a lot of hard work are necessary to find a model description of the multiparticle forces in an atom or a nucleus that is in agreement with experiment. Further, different atoms or different nuclei require different models. Experimental physicists, on the other hand, use as much if not more of the same intuition, ingenuity, and long hours to test the models and to obtain data that will form the basis for new models.

If a particle is a localizable glob of energy which also behaves like a wave, it can be represented by a wave packet as we saw in Figure 18-1. The **uncertainty principle**, enunciated by Werner Heisenberg in 1927, must always be a natural consequence of *any* wave motion which is also localizable. This principle and its physical foundations have often been misinterpreted by philosophers and laymen. We will show here that, if particles have wave properties or if waves are localized, they must always obey the uncertainty principle.

Heisenberg states this even more clearly in the following passage:

The concepts of velocity, energy, etc., have been developed from simple experiments with common objects, in which the mechanical behavior of macroscopic bodies can be described by the use of such words. These same concepts have then been carried over to the electron, since in certain fundamental experiments electrons show a mechanical behavior like that of the objects of common experience . . . . According to Bohr, the processes of atomic physics can be visualized equally well in terms of waves or particles. Thus the statement that the position of an electron is known to within a certain accuracy Δx at the time t can be visualized by the picture of a wave packet in the proper position with an approximate extension Δx. By "wave packet" is meant a wavelike disturbance whose amplitude is appreciably different from zero only in a bounded region. This region is, in general, in motion and also changes its size and shape, i.e., the disturbance spreads. The velocity of the electron corresponds to that of the wave packet, but this latter cannot be exactly defined because of the diffusion which takes place. This indeterminateness is to be considered as an essential characteristic of the electron, and not as evidence of the inapplicability of the wave picture . . . . From the simplest laws of optics, together with the empirically established law $\lambda = h/p$, it can readily be shown that

$$\Delta x\, \Delta p \geq h$$

frequency   planks constant   momentum

This uncertainty relation specifies the limits within which the particle picture can be applied. Any use of the words "position" and "velocity" with an accuracy exceeding that given by [the] equation is just as meaningless as the use of words whose sense is not defined.[1]

Werner Heisenberg in 1933.

[1] Werner Heisenberg, *The Physical Principles of the Quantum Theory* (Chicago: The University of Chicago Press). Copyright © 1930, p. 10. Used by permission.

We will now derive the uncertainty principle. Consider a train of waves in the form of a reasonably localized packet (see Figure 18-4). The wave train is characterized by a wavelength λ and by a distance Δx over which it has nonzero amplitude. This wave packet is localized in the region Δx and does not exist outside of that region. It is clear from Figure 18-4 that the position of the particle-wave is no longer well defined, but is uncertain by the length of the wave train Δx (Δ here means "uncertainty in"). A wave packet like this can only be constructed from the sum of a large number of infinite sine waves. (A sine wave can be thought of as a periodic wave which has a constant wavelength and constant amplitude.) This idea may be demonstrated by super-positioning two sine waves of slightly different wavelengths. Over the length of the wave packet, wave (1) must contain *at least* one more cycle than wave (2) so that the two waves will be in phase near the middle of the wave packet but out of phase and canceling one another at both ends of the wave packet of length Δx (see Figure 18-5). This means that if there are N wavelengths λ_2 in Δx, there must be *at least one* more or $(N + 1)$ wavelengths λ_1 in the same length Δx:

$$\frac{\Delta x}{\lambda_2} = N$$

and

$$\frac{\Delta x}{\lambda_1} \geq N + 1$$

Subtracting these two equations then yields

$$\frac{\Delta x}{\lambda_1} - \frac{\Delta x}{\lambda_2} \geq N + 1 - N = 1$$

or

$$\Delta x \left(\frac{1}{\lambda_1} - \frac{1}{\lambda_2} \right) \geq 1$$

If we multiply both sides by h and define the uncertainty in momentum p via de Broglie's relation $p = h/\lambda$

$$\frac{h}{\lambda_1} - \frac{h}{\lambda_2} = \Delta p = p_1 - p_2$$

we arrive at Heisenberg's uncertainty principle:

$$\Delta x \, \Delta p \geq h \tag{18-1}$$

which states that the uncertainty in position times the uncertainty in momentum is always larger than the constant h.

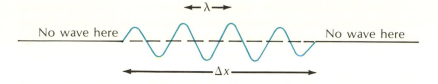

18-4 A wave packet localized
in the region Δx.

This relationship has rather deep physical as well as philosophical implications. Physically, it tells us that we cannot know
both the position and momentum of a particle to an arbitrary
degree of precision. If, for example, we know the precise location
of a particle ($\Delta x = 0$), the uncertainty principle tells us that we
can have no knowledge of the particle's momentum (or its velocity) since Δp must be infinite.

The uncertainty principle also changes the deterministic picture given by Newtonian physics. Classical mechanics states that
if we knew the exact positions and momenta of all the particles
in the universe at some time, we could, at least in principle,
calculate the past and future course of the universe from the exact
laws of physics. Such reasoning—that the universe itself is like
a machine—leads to problems concerning the effect of free will
on the actions and decisions we make. The uncertainty principle,
however, tells us that we cannot have this precise information
at any time. Thus, the best we can do is to give a statistical
prediction of the probability of future events and expect such
predictions to become worse or more uncertain as time evolves
from the present.

Another consequence of the uncertainty principle is the observer's effect on the event observed. We see or measure the
location of objects because a light source emits a stream of pho-

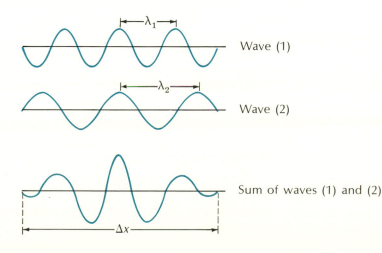

Wave (1)

Wave (2)

Sum of waves (1) and (2)

18-5 Two sine waves of
different wavelengths are added.

tons which bounce off the object we observe into our eyes or our measuring apparatus. The energy or mass of the light which bounces off a macroscopic object is much smaller than the mass of the object itself, so the total observation process (which includes the light particles as well as our eyes and brain) does not interfere appreciably with the motion of the object we observe.

A physicist in a laboratory experimenting on an atomic or nuclear system "sees" in an analogous fashion. He directs a beam of particles (photons, electrons, protons) at a target and observes, with the help of electronics in varying degrees of sophistication, the pattern made by the scattered particles. As we discussed in the previous section, a physicist sees into the submicroscopic world by interpreting these patterns for the forces involved via models. However, the wavelength of the beam particles must be at least as small as the size of any detail we want to observe. Unfortunately, small wavelengths only occur with large momenta (hence, high energy), as de Broglie showed in his formula $\lambda = h/p$. The experimental study of the details of an atom or nucleus requires beam particles whose energies are comparable to the energy of the system to be studied. In this way, the very fact that a microscopic object has been observed will affect the result of the observation.

For an analogy, suppose we wished to trace the flight of a basketball by throwing baseballs at it and observing the resulting pattern of scattered baseballs. Clearly, each collision would alter the trajectory of the basketball appreciably. This is precisely the situation in quantum mechanics. If we wish to locate an electron, we must use photons of very short wavelength—about the size of the electron. Every time one of these photons bounces off the electron, the electron will recoil from the collision and, although it is localized, we will have no exact knowledge of its momentum after each collision. We could of course use less energetic photons which would not transfer appreciable momentum to the electron. Unfortunately, this would require using very long-wavelength photons which, in turn, would not tell us much about the electron's position.

This situation seems truly paradoxical. The uncertainty in position Δx and the uncertainty in momentum Δp are intermingled because of the wave nature of matter. The uncertainty principle tells us that the degree of this intermingling is

$$\Delta x \, \Delta p \geq h$$

The inescapable conclusion is that the detection apparatus cannot, *even in principle,* be divorced from the event to be observed. This effect, although true on all levels of observation, becomes most crucial on an atomic-nuclear scale.

In Chapter 17 (pages 228–32) we discussed the bound states that exist for waves which are confined at both ends by a force, like the waves on a guitar string fixed firmly at both ends. For a string of length L we found a succession of different wavelengths $\lambda = 2L$, L, $\frac{2}{3}L$, $\frac{1}{2}L$, and so on for all of the possible allowed vibrations. If an electron in an atom is under similar conditions (i.e., confined between two "walls" or in a **force well** of width L), then it will exhibit wavelike qualities and have any one of a succession of the same wavelengths: $\lambda = 2L$, L, $\frac{2}{3}L$, $\frac{1}{2}L$, or, in general, $\lambda = 2L/n$ where n is any whole number. The energy of the electron bound in the force well is simply the kinetic energy $E = \frac{1}{2}mv^2 = p^2/2m$, where p is the momentum of the electron as it moves between the walls and is related to the wavelength by de Broglie's formula $p = h/\lambda$. Thus, the energy is

$$E = \frac{h^2}{2m\lambda^2} \qquad \text{(18-2)}$$

This is a very general result and does not depend on the specific shape of the force or the well that contains the electron. For the **square well** in the example above, $\lambda = 2L/n$ ($n = 1,2,3,4,\ldots$); therefore the energies $E = h^2n^2/8mL^2$ are the only possible energy values for the electron. These energy values are **quantized**; by this we mean that only certain discrete values are allowed. If we give $h^2/8mL^2$ the symbol E_1, then the possible energies of the electron in the square well are E_1, $4E_1$, $9E_1$, $16E_1$, and so on (see Figure 18-6). They increase as the square of the quantum number n. It should be emphasized that these energies and only these energies are allowed for the electron.

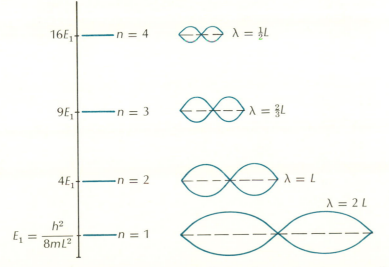

18-6 The energies and wave pictures for an electron in a square well.

Not all physical situations can be approximated by the model of the square well. In the hydrogen atom, for example, the electron is bound to the proton by the attraction of the Coulomb force—a $1/r^2$ force. Unlike the energy levels for an electron in a square well, the energies for an electron in the hydrogen atom are measured negatively from zero for convenience (see Figure 18-7). We see that different forces on the electron produce different energy levels. The square well levels in Figure 18-6 are not at all like the hydrogen levels in Figure 18-7: The energy values and the spacing between levels are completely different; however, in both cases the electron can have any one of a number of discrete energy values associated with the quantum number n, where n is a whole number (an integer). This always holds true, no matter what the force by which the electron is bound looks like. Furthermore, when the electron "jumps" from a higher energy level to a lower one, the difference in energy E between the two levels always manifests itself in the form of a photon whose frequency is given by the Planck formula $E = hf$.

In complex, multielectron atoms heavier than hydrogen, the force that binds the electron is much more complicated than a simple Coulomb force since there are many other charged particles around (other electrons and the nucleus), each exerting a force on the electron being studied. The sum of all these forces is generally too complicated mathematically to express exactly in a simple picture, as we discussed in Section 18-3. As in the case of the scattering problem, the theorist's job here is to find a simple and solvable model force well that approximates the actual force. He places an electron in the "well" and solves the Schrödinger equation for the model force to find the energy levels.

Thus far we have restricted our discussion to the energy levels of an electron wave moving in only one dimension. Real electrons,

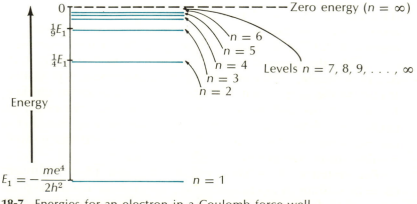

18-7 Energies for an electron in a Coulomb force well.

however, move in three dimensions and are contained or bound to an atom by forces which exert their attraction in all three space dimensions. Since the motion of the electron in one direction (say, along the x axis) is independent of its motion along a perpendicular direction (say, the y axis), it should be reasonable to assume that there must be a separate, independent quantum number for each of the three dimensions (the x, y, and z axes) in the real world. Thus the energy-level diagrams for a real electron are considerably more complicated than the one-dimensional model discussed here.

18-6

Multielectron Atoms

The more interesting, but also the more complicated case of many electrons being bound to one nucleus occurs in all atoms heavier than the hydrogen atom. In each of these atoms, the number of electrons exactly equals the number of positive charges in the nucleus, so that the atom is electrically neutral when viewed from the outside. However, the individual electrons are moving in a very complicated force due to the presence of the nucleus as well as all of the other electrons. Each electron has quantum numbers associated with its motion which determine its energy. Each moves in the space around the spherical nucleus, governed by its own quantum numbers even when other electrons are present.

The **atomic number** Z is simply equal to the number of positive charges in a nucleus or, alternatively, to the number of orbital electrons in an atom. In the lightest element, hydrogen (H), there is only one electron. It is normally found in the ground state and is designated by the lowest possible value of the quantum number $n = 1$. This is the lowest energy "orbit" the electron can occupy. Its negative charge exactly equals the single positive charge on the nucleus of hydrogen (see Figure 18-8). Thus we say that hydrogen has an atomic number $Z = 1$.

Two electrons "orbit" the doubly-charged nucleus of helium (He). This element is more complicated than hydrogen in that we must determine the quantum numbers for two electrons. We assume that helium's first electron is in the ground state with $n = 1$, like the electron in the hydrogen atom. But what is the quantum number of the second electron and where is it located? It may be the same as the first electron, but this seems unlikely. If *all* electrons had the same quantum number, we could not explain the differences in chemical properties, valences, and so on, exhibited by the various elements, nor could the periodic regularities of Mendeleev's table be explained.

For instance, it is known that hydrogen $(Z = 1)$, lithium $(Z = 3)$, sodium $(Z = 11)$, and so on, have similar chemical properties; they form a **family** whose members are all very active

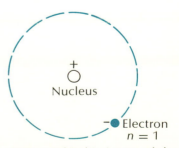

18-8 "Orbital" picture of the hydrogen atom.

chemically. Florine ($Z = 9$), chlorine ($Z = 17$), and bromine ($Z = 35$) form another family with their own similar chemical properties; these are also very chemically active elements that combine readily with members of the hydrogen-lithium-sodium family to form very stable molecules like sodium chloride and hydrogen chloride. On the other hand, the noble gases helium ($Z = 2$), neon ($Z = 10$), argon ($Z = 18$), and so on, form a third and very different family that exhibits little chemical activity; the atoms of these elements do not easily form molecules with atoms of other elements.

18-7

The Pauli Exclusion Principle and Electron Spin

These experimental facts on the *periodicity* (with increasing Z) of chemical properties led Mendeleev to propose his periodic chart (Figure 17-2) in 1869. With the discovery of quantum mechanics, these same facts led to the **exclusion principle** proposed by Wolfgang Pauli in 1925. This principle states that no two electrons in an atom may have the same quantum number, the quantum equivalent of the familiar classical statement that no more than one object can occupy the same space at the same time. Pauli's exclusion principle, as we will see below, automatically produces the chemical periodicity Mendeleev observed since it allows the electrons in an atom to successively fill **shells** corresponding to the quantum number n (called the radial quantum number) as the atomic number Z increases.

Thus for helium, with an atomic number of $Z = 2$, this rule would allow one electron to be in $n = 1$ and the second electron to orbit in the next highest shell with $n = 2$. But this picture (see Figure 18-9) is incorrect because it indicates that helium is similar to hydrogen, since it shows a single electron in the $n = 2$ shell with the $n = 1$ shell closed. Experimentally, however, helium is known to be a noble gas and should therefore be characterized by a completely **closed-shell** picture. Furthermore, it is known that the next heaviest atom, lithium ($Z = 3$) with three electrons, is

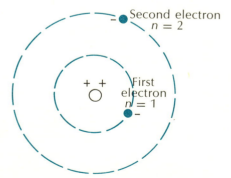

Second electron
$n = 2$

First electron
$n = 1$

18-9 An incorrect orbital picture of the helium atom.

doesn't take spin into account

ref next page

PAULI

Wolfgang Pauli was born in 1900 in Vienna and grew up in a villa on the outskirts of the city. He finished his university training at Munich at the age of 19, and astonished the world of physics by writing one of the clearest and best expositions of Einstein's relativity theory when he was only 20. His work, published in the German Encyclopedia of Physics, won much fame for its penetrating insight. □ Pauli then traveled to Göttingen in Northern Germany and, in 1922, came to Bohr's Institute in Copenhagen. Pauli threw himself into his work with a great deal of both mental and physical energy. He would often rock back and forth in his chair while he thought, oblivious of anyone around him, until the room literally shook with his movements. He was disdainful and suspicious of experimental physics. The "Pauli Effect," as it has been jokingly called, refers to Pauli's awkwardness in a laboratory where glassware or some other apparatus would often break spontaneously when he entered. Once a mysterious explosion destroyed some vacuum equipment in the Göttingen laboratory at the precise moment Pauli's train pulled into the station. □ Pauli was said to have resembled the Buddha in both his rotund physical appearance and his enigmatic approach to life. He strongly criticized his colleagues, who benefited from his keen mind even though they may not have cared for his sharp, caustic tongue. □ Although Pauli went to Hamburg after only a year at Copenhagen, he remained a familiar figure at Bohr's Institute during the exciting years after 1924 when the concept of quantum mechanics was being developed. Many physicists—Heisenberg, Bohr, and Schrödinger among them—profited greatly by their talks with Pauli there. □ During the war, Pauli came to the United States to work with Dirac and Einstein in the Princeton Institute for Advanced Studies. The Pauli exclusion principle, one of Pauli's many contributions to theoretical physics, won him the Nobel Prize in 1945. Pauli became a naturalized American citizen in 1946 and continued his work in physics until his death in 1958.

Wolfgang Pauli with Einstein.

The Niels Bohr Library, American Institute of Physics

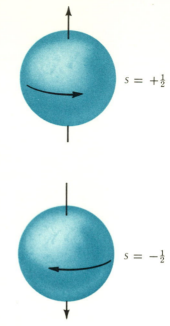

$s = +\frac{1}{2}$

$s = -\frac{1}{2}$

18-10 Semiclassical picture of electron spin.

the closest element with chemical properties similar to those of hydrogen. Nature is telling us something, but what?

To solve this dilemma, in 1925 two theoretical physicists, George Uhlenbeck and Samuel Goudsmit, proposed that not one but *two* electrons could fit into each atomic orbit for the same value of n. To reconcile this proposal with the Pauli exclusion principle which demands that every electron in an atom must have its own quantum number different from that of any other electron, Goudsmit and Uhlenbeck proposed a new quantum number for each electron which could have only one of *two* possible values *independent of* the values of n. They called this new quantum number **spin** and proposed that it was an intrinsic property of any electron, whether that electron was in an atom or not. The classical picture we can construct from this is an electron spinning on its own axis like a top. Since the electron has only two spin values, it can spin clockwise (angular momentum or spin vector pointing *down*) or counterclockwise (spin *up*), as in Figure 18-10.

In accordance with the rules for quantum numbers we have learned about thus far, any quantum number can change from one value to another in integer steps only. Since the quantum number s for spin has been given only two values, they are represented by $s = +\frac{1}{2}$ and $s = -\frac{1}{2}$ so that the change in s is only one unit and s has symmetric values about zero. Now we can see how (1) the Pauli exclusion principle and (2) the electron spin allow us to build up a model for the periodic table that is in agreement with nature.

For helium ($Z = 2$), the first electron has the quantum number $n = 1$ and a spin quantum number s equal to either $+\frac{1}{2}$ or $-\frac{1}{2}$. (Let us assume here that $s = +\frac{1}{2}$.) The second electron then "orbits" in the same shell ($n = 1$) with quantum numbers $n = 1$, $s = -\frac{1}{2}$ (Figure 18-11). This satisfies the Pauli principle (since the quantum numbers for the electrons are not the same because their spin orientations differ) and produces a closed $n = 1$ shell picture of the helium atom, thus explaining its chemical property of being a noble or an inactive gas. A closed shell of electrons is a very stable configuration which does not change or combine with other atoms easily.

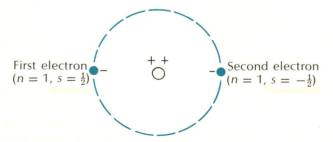

First electron
($n = 1$, $s = \frac{1}{2}$)

Second electron
($n = 1$, $s = -\frac{1}{2}$)

18-11 Correct orbital picture of the helium atom.

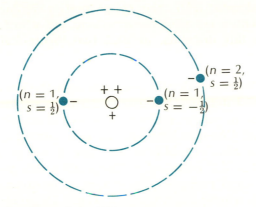

18-12 Orbital picture of the lithium atom.

For lithium (Li, $Z = 3$), the first two electrons completely fill the $n = 1$ shell as in the helium atom; the third electron must then orbit the $n = 2$ shell, making its quantum numbers $n = 2, s = +\frac{1}{2}$ (Figure 18-12). Since this configuration looks just like the one for the hydrogen atom, except that lithium is in the $n = 2$ rather than the $n = 1$ shell, we expect Li to behave in a manner chemically similar to H, which it does.

For atoms heavier than lithium, the situation becomes more complicated mathematically due to the three-dimensional nature of the problem. Briefly, each electron actually has *three* quantum numbers assigned to it in addition to the spin quantum number s. One of these three is the radial quantum number n we have already discussed. For the cases of $Z = 1, 2,$ and 3, the other two quantum numbers (the angular momentum l and the magnetic quantum number m) are unimportant since they each equal zero. But in heavier atoms where $n = 2, 3, 4,$ and so on, these other two quantum numbers have nonzero values. Consequently, for $n = 2$, there are *four* possible combinations of l- and m-values in addition to *two* possible spin values for *each* electron. Thus, for example, it takes eight electrons (4×2) to completely fill up the $n = 2$ shell and the next closed-shell element must occur at $Z = 10$. At this "magic" number 10, the $n = 1$ shell is closed with two electrons and the $n = 2$ shell is closed with eight electrons in an atom of the noble gas neon (Ne).

The possible values of l and m are predicted by the quantum description of the atom. In this way, with the use of (1) quantum mechanics via the Schrödinger equation, (2) the Pauli exclusion principle, and (3) the electron spin, the systematics and chemical properties of the elements are explained. Detailed calculations that predict the properties of many electron atoms and molecules today are all based on further simplifying models and assumptions. Such models are necessary due to the vast mathematical complications that arise when we try to perform these calculations with three or more particles.

18-9
More Philosophy

In Section 18-4 we discussed some of the implications of the uncertainty principle and how it changed the deterministic picture of measurement by providing the observer or observing apparatus with an important role in the observation process. In relativity, what is observed depends upon the fact that it is being observed or upon the state of motion of the observer. We see that both relativity and quantum mechanics give new roles to the observer that were not present in classical Newtonian physics. Is nature trying to tell us something again? Is there a more fundamental principle that unifies these two roles of the observer? No one knows at present. We close this chapter with a few more general comments.

We have been presented with a new concept—quantization. Classically, any physical quantity such as energy (and charge) could be subdivided as finely as necessary, but now quantum mechanics yields an entirely new structure. If the submicroscopic world is investigated, a granular, rather than a continuous, structure is found: The smallest charge is that of an electron; only discrete energies occur in a bound state system and all other energies are forbidden. When an experiment is performed, certain directions in space are quantized. Light consists of discrete energy packets—photons, and so on, and so on. But what else is quantized? Thus far, both time and distance are still believed to be continuous in the sense that an elementary quantum or "grain" of either has yet to be found experimentally or required theoretically. But even this is not a closed subject, as we will see in Chapter 20.

Quantum mechanics has also led to the abandonment of two classical concepts of scientific theory—cause and effect relations and determinism. Newtonian mechanics tells us that we can predict the outcome of a given experiment exactly if we have the initial data. Quantum mechanics only discusses the probability that a given event occurs. Even with the best possible experimental arrangement, it is impossible to predict exactly what will happen; only the odds are predictable, not the precise outcome of nature—truly a major change from the old "Newtonian" understanding of nature.

It is interesting to note that Einstein, who contributed immensely to quantum theory himself, did not regard statistical interpretation as the final word:

The de Broglie–Schrödinger wave fields were not to be interpreted as a mathematical description of how an event actually takes place in time and space, though, of course, they have reference to such an event. Rather they are a mathematical description of what we can actually know about the system. They serve only to make statistical statements and predictions of the results of all measurements which we can carry out upon the system . . . .

The aim of the theory is to determine the probability of the results of measurement upon a system at a given time. On the other hand, it makes no attempt to give a mathematical representation of what is actually present or goes on in space and time. On this point the quantum theory of today differs fundamentally from all previous theories of physics, mechanistic as well as field theories. Instead of a model description of actual space-time events, it gives the probability distributions for possible measurements as functions of time.

It must be admitted that the new theoretical conception owes its origin not to any flight of fancy but to the compelling force of the facts of experience. All attempts to represent the particle matter, by direct recourse to a space-time model, have so far ended in failure. And Heisenberg has convincingly shown, from an empirical point of view, that any decision as to a rigorously deterministic structure of nature is definitely ruled out, because of the atomistic structure of our experimental apparatus. Thus it is probably out of the question that any future knowledge can compel physics again to relinquish our present statistical theoretical foundation in favor of a deterministic one which would deal directly with physical reality . . . .

Some physicists, among them myself, cannot believe that we must abandon, actually and forever, the idea of direct representation of physical reality in space and time; or that we must accept the view that events in nature are analogous to a game of chance. It is open to every man to choose the direction of his striving; and also every man may draw comfort from Lessing's fine saying, that the search for truth is more precious than its possession.[2]

Einstein spent the last three decades of his life searching for a deterministic, non-statistical model of quantum mechanics. But his "search for truth" was unsuccessful.

1 What is meant by a "bound state"? What is its importance?

2 How does quantum mechanics explain the success of de Broglie's matter-wave theory?

3 How do physicists interpret the meaning of the "wave function" of quantum mechanics? How is their interpretation a departure from the determinism of classical physics?

[2] Albert Einstein, *Ideas and Opinions* (New York: Crown Publishers, Inc.). Copyright © 1954, p. 332. Used by permission of Crown Publishers, Inc.

4 State Heisenberg's uncertainty principle. To what sort of uncertainty does it apply?

5 Do you believe that a logic could exist which *in principle* man could not understand? Could laws of nature exist which man could not perceive or discover?

6 The laws of Newtonian mechanics substantiated or predicted the entire past, present, and future history of the universe. Relativity changed the specific form of these laws, but the determinism remained. Given the uncertainty principle, does the new quantum mechanics still present a picture of the universe in which man's free will plays a decisive role? Explain.

7 Are topics like "free will" proper subjects for discussion in terms of the laws of physics? Do physicists have the right to assume that the laws of physics are meaningful is discussions of such topics? Explain.

8 What do we mean by the "energy levels" of an atom? Why do we say that the energy of an atom is "quantized"?

9 What are "quantum numbers"? Why are they useful in describing the properties of atoms of elementary particles?

10 Atoms have been described as "miniature solar systems." Explain why this description is partly correct and partly incorrect.

11 What evidence led to the hypothesis that a massive, positively-charged nucleus was at the center of an atom?

12 Why do atoms have more than one characteristic frequency of emitted light?

13 State the Pauli exclusion principle. Why is it important in atomic structure? How would atoms behave if no exclusion principle existed?

14 How does our knowledge of atomic physics help to explain Einstein's "work function" in the photoelectric effect?

15 For atoms with very many electrons, the calculation of energy levels becomes very difficult and it is hard to predict experimental results. Do you think this is a failure of the basic theory of atomic structure or simply a manifestation of the complexity of the problem? Explain.

16 Do you think that scientific knowledge has a greater impact on the personal lives of people today in this world of quantum mechanics and the resultant technology than it did in the sixteenth and seventeenth centuries? Explain.

Questions Requiring Outside Reading

1 Describe the modern concept of the atom. In what ways does this concept differ from previous theories?

2 Starlight is known to consist of very bright "lines" of certain wavelengths. Can you think of a way in which atomic physics might explain the existence of these bright lines?

3 The first great synthesis in physics—that of Copernicus, Galileo,

Kepler, Newton, *et al.*—had very profound political effects which resulted from the challenge it presented to the authority of the Catholic church. What are some of the important political questions associated with scientific research today in connection with the quantum revolution? Is the political impact of science as important now as it was in the Renaissance? Explain.

4 Give some examples of applications of the uncertainty principle.

5 What interaction occurred in the 1920's between Heisenberg, Schrödinger, de Broglie, and Bohr?

6 How did the probability interpretation of quantum mechanics evolve?

7 How did Einstein view the probability interpretation?

8 If matter really consists of waves, why are ordinary macroscopic objects (such as billiard balls or baseballs) localizable and why do they seem to have definite trajectories?

1 Atoms obey conservation of energy when they radiate light. Describe how this radiation of light takes place. (A diagram might be helpful.)

Problems

2 Why do atoms emit only certain discrete frequencies of light and not others?

3 A bowling ball has a mass of 7×10^3 g and is seen to be moving with a speed of 1000 ± 2 cm/sec. What is the uncertainty in its momentum? What is the smallest possible uncertainty in its position?

4 Explain in physical terms why the observed state of a quantum mechanical system should be different from the unobserved state.

5 Calculate the first several energy levels for an electron with a mass of 9.1×10^{-27} g in a one-dimensional square well. Assume the length L of the well is approximately the size of a light atom like hydrogen ($L = 10^{-7}$ cm).

6 What is the energy of the light photon emitted when an electron jumps from level b, c, or d to the ground state level a in the square-well problem above? (*Hint:* Refer to Figure 18-6, page 245.)

7 What are the frequency and the wavelength of the photon in Problem 6?

8 Calculate the numerical values for the first several energy levels in the hydrogen atom and compare these values with those you calculated in Problem 5.

9 Because of its spin, the electron behaves like a tiny bar magnet. Would you expect two electrons to have a lower energy (their more probable state) when their spins are aligned in the same direction or in opposite directions?

10 What is a "closed-shell" atom? Explain how it behaves chemically in terms of the values of the quantum numbers of its electrons.

19

Nuclear Physics—An Application of Quantum Mechanics

Although the concept of an atomic nucleus did not evolve until Ernest Rutherford's experiments in 1911, a large body of very interesting, if puzzling, information about the interior of the atom had already accumulated at the end of the nineteenth century. By 1896, Henri Becquerel had identified the following energetic radiations emitted from uranium salts:

Types of Nuclear Radioactivity

Radiation	Description	Charge	Rest Mass (in electron masses)
(α) alpha	helium nucleus	positive	about 4000
(β) beta	electron	negative	1
(γ) gamma	energetic photon	none	0

The two unusual features of this nuclear **radioactivity** were (1) there were three different types and (2) the energies were much *higher* than the energies previously ascribed to atomic processes. Becquerel's classifications were to lead directly to the Curies' discovery of two radioactive elements, polonium and radium, as decay products of uranium. But could these energetic radiations come from *atomic* processes, or was there more inside of atoms than was previously suspected?

Rutherford was the one who set nuclear physics on its proper course when he demonstrated experimentally that nearly all of the mass of an atom is contained in a very small volume at its center. The exact size and nature of this nucleus was unimportant in *atomic* physics. All that was necessary to establish the atomic theory was the knowledge that the nucleus could be considered as a point charge to a good approximation. Its exact dimensions were unimportant as long as they were small in comparison with the radii of the electron orbits (about 10^{-8} cm for hydrogen). Rutherford and other experimenters with nuclear scattering who followed him were able to demonstrate that the inverse square law of Coulomb repulsion held for the nuclear charge to within a distance of 10^{-12} cm. (It is now known that even the largest nucleus—uranium—has a radius no greater than 10^{-12} cm.) Perhaps Rutherford's nucleus was the source of the radiations Becquerel observed.

19-2

What Is in the Nucleus?

Rutherford's model was accepted and, until early in the 1920's, physicists believed that the nucleus itself was a mixture of positive **protons** and negative **electrons**, but fewer electrons than protons so that the nuclear charge would always be positive. This model seemed reasonable enough; beta rays, which are simply energetic electrons, were one type of nuclear radioactivity. Because these electrons were emitted by a certain nuclei, it was concluded that they must have previously existed within the nucleus. But with the dawn of quantum mechanics and the uncertainty principle, it became apparent that a particle with the small mass of an electron would have such a long wavelength that it could not possibly exist in so small a volume as a nucleus. Then what did the nucleus contain besides protons? Rutherford and James Chadwick advanced the hypothesis that an electrically-neutral particle about the size of a proton was the uncharged constituent of the nucleus. Because of its electrical neutrality, this particle escaped detection until 1932 when Chadwick discovered the **neutron**.

Since nearly all of the mass of an atom is contained in the nucleus (electrons contribute very little to the atomic mass), one immediate deduction is that "**nuclear matter**" has no counterpart in our normal experience; it is unbelievably dense, about a billion tons per cubic inch. Evidently the nucleus is something new in our experience and we must be wary of trying to ascribe to it "common-sense" properties characteristic of matter that we experience in everyday life.

One striking feature of the nucleus is that extremely high energies are required to keep it together. Unlike the atom, the nucleus is not electrostatically neutral but consists of anywhere from one to 103 positively-charged protons mixed in with an approximately equal number of neutral particles called neutrons, thus creating a fantastically strong Coulomb repulsion which is constantly trying to tear the nucleus apart. But some powerful attractive force (which must be of an extremely short range because it does not affect Rutherford's alpha particles and which must be much stronger than the Coulomb force) holds the nucleus together. For some peculiar reason, this **nuclear force** does not affect electrons at all. A completely satisfactory explanation of this strong force has yet to be found, and its known properties are being carefully studied in laboratories around the world today.

19-3

Binding Energy

The strongest known force exists between neutrons and protons in a nucleus. The helium nucleus, for instance, consists of two protons and two neutrons. The attractive force between these four particles is far stronger than the repulsive Coulomb force between the two positively-charged protons, since the helium nucleus, an alpha particle, is extraordinarily stable. It would require 28 million electron volts (**MeV**) of energy (1 MeV = 1.6×10^{-6} ergs) to break up this nucleus into its four constituent particles. We call this the **binding energy** of the nucleus. This energy is simply the difference between the mass of, in this particular case, the helium nucleus and the sum of the masses of the four individual constituent particles, using Einstein's mass-energy equation $E = mc^2$.

Example 19-1:

Deuterium is the heavy isotope of hydrogen and its nucleus, the deuteron, consists of one proton and one neutron. What is the binding energy of the deuteron if its mass is 3.344×10^{-24} g, the mass of a free proton is 1.673×10^{-24} g, and the mass of a free neutron is 1.675×10^{-24} g. The sum of the proton and neutron masses is 3.348×10^{-24} g. Subtract the deuteron mass of 3.344×10^{-24} g from this to obtain the mass difference of 0.004×10^{-24} g = 4.0×10^{-27} g. Then, use $E = mc^2$ to find the binding energy given by this mass difference:

$$E = 4.0 \times 10^{-27} \times 9 \times 10^{20} = 3.6 \times 10^{-6} \text{ ergs} = \textbf{2.2 MeV}$$

The binding energy of a nucleus is the energy required to break it up into all of its constituent neutrons and protons, collectively called **nucleons**. Conversely, it is also the energy that would be released if the separate constituent nucleons were to form the nucleus. In Example 19-1 above, a separate neutron would combine with a proton to form the **deuteron**, thereby releasing 2.2 MeV of energy. This process, the building of heavier nuclei from their constituent parts or from lighter nuclei, is called **nuclear fusion**.

19-1 Energy-level diagrams for the deuteron and for the helium nucleus.

It is interesting to study the nuclear force—the force between the neutron and proton—in the deuteron. The deuteron system in the nuclear case is analogous to the hydrogen atom in the atomic case (although deuteron energy is about 100,000 times hydrogen energy). Each system has only two particles and is thereby the simplest to analyze in its respective system. The simplicity of the hydrogen atom and the regularity of its spectral lines gave Bohr and de Broglie the clues which enabled them to make the breakthrough in atomic structure. Likewise, the deuteron provided analogous clues which have led to a more complete understanding of nuclear structure. One interesting comparative fact is that the deuteron has only one bound state ($n = 1$), whereas the hydrogen atom has an infinite number of bound states ($n = 1,2,3, \ldots$). The reason for this is that the nuclear force in the deuteron has a very **short range** (it falls to zero over a very short distance), while the Coulomb force that binds the hydrogen atom is a $1/r^2$ force that is still effective a very long distance away. Consequently, there are wave functions for hydrogen that correspond to the electron bound a long distance away (large n) from the proton. In a similar manner, the other very light nuclei like helium also have only one bound state. As we have seen, in the deuteron the energy level is at -2.2 MeV and the energy level in the helium nucleus is at -28 MeV, as shown in Figure 19-1.

In heavier nuclei with more than five nucleons there are many energy levels and their arrangement or spacing is usually complicated, as shown in Figure 19-2 for the oxygen nucleus. Nuclear physicists today are primarily occupied with trying to measure and understand the complicated energy-level structures of all of the nuclei of each element.

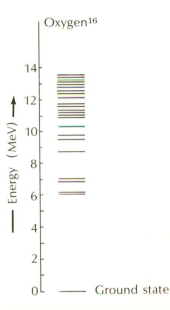

19-2 Energy-level diagram for the oxygen nucleus for all levels below 14 MeV.

Soon after Chadwick's discovery of the neutron, Enrico Fermi in Rome began to send neutron missiles into the interior of various nuclei, reasoning that neutrons would enter the nucleus more readily than protons because the neutrons would not be repulsed by the Coulomb force which stops most charged particles. In the

19-4

Nuclear Fission

process of methodically bombarding many known elements with neutrons, Fermi artificially created a number of new nuclei which he called **isotopes** (from the Greek "iso topos," meaning "the same place" and referring to the periodic table of the elements). The only difference between a given element and one of its isotopes was the number of neutrons in the nucleus. Both were chemically identical since they had the same electron structure (which is what determines an atom's chemical properties). In certain elements more than one isotope occurred naturally, accounting in large part for the fractional atomic weights of elements in the periodic table.

By the late 1930's, many experimenters were bombarding nuclei with neutrons. The question arose as to whether it would be possible to create a brand new element at the heavy end of the periodic table by bombarding the heaviest known element—uranium—with neutrons. The attempt produced some startling results. In 1939 two German chemists, Otto Hahn and Fritz Strassman, made chemical studies of the products which resulted from the slow neutron bombardment of uranium. Among the reaction products they were astonished to find barium, an element with a nucleus about half the size of the uranium nucleus. How had pure uranium produced barium?

Lise Meitner and Otto Frisch, refugees from Nazi Germany who were working in Stockholm and Copenhagen, pointed out that the presence of barium as a product of the neutron-uranium reaction meant that the uranium nucleus was undergoing a splitting or **fission** process. This idea initiated feverish research on the neutron-uranium reaction by physicists everywhere. Within a few months, it was determined that a naturally-occurring isotope of uranium of weight 235 (U-235 meaning 235 nucleons in the nucleus) was the fissioning nucleus.

Fission is the splitting of a heavy nucleus into two parts or fission fragments of comparable, but not necessarily equal, size (see Figure 19-3). Fission can occur spontaneously in a few of the heaviest nuclei, but it can also be greatly stimulated by the provision of a little extra energy in the form of bombarding neutrons. The energy released when fission takes place is very large—an average of 200 million electron volts (MeV) per fission. Most of

Neutron

19-3 An illustration of fission using the "Liquid Drop Model."

FERMI

Enrico Fermi was born in 1901 in Rome and was educated in the public schools there. As a young child, he delighted in improvisation and was gifted in building things out of whatever materials were available to him. In 1918, he won a scholarship to the Normal School in Pisa, where he obtained his doctoral degree in 1922. He continued his postdoctoral studies in Germany, and returned to Italy at the age of 25 to become a full professor at the University of Rome. There he was the acknowledged leader ("the pope") of an active and productive group of several young physicists who were working on the new quantum physics in Italy. □ Fermi's work at the university on neutron absorption by heavy elements won him the Nobel Prize in 1938. Since life in Mussolini's Italy was becoming increasingly uncomfortable (Fermi's wife had Jewish relatives), he traveled to Sweden with his family to receive the prize with no intention of returning to Italy. After the Nobel ceremonies, the family emigrated to America and Fermi began teaching at Columbia University the following year. □ In 1942, Fermi left Columbia for the University of Chicago, where he led the group that discovered the first self-sustaining chain reaction. With the knowledge that uranium would sustain a chain reaction, the nuclear bomb became a realistic possibility. □ Fermi was simple in his ways and was able to accept other people as they were without trying to manipulate them. He was a warm person, modest and friendly. In physics, simple theories that could be easily grasped appealed to him. His lectures were extraordinarily clear and were always easily understood by students, professionals, and laymen alike. His influence on the younger physicists at the Institute for Nuclear Studies at the University of Chicago was immeasurable, and his death at the early age of 54 was mourned deeply by both students and colleagues—indeed by the whole world of physics.

Enrico Fermi

The Niels Bohr Library, American Institute of Physics

it is the kinetic energy (essentially heat) of the separating fragments, although a lot of electromagnetic radiation is also released. The difference between the greater combined binding energy of the two fission fragments and the lesser binding energy of the original uranium nucleus is the energy of fission. Another way of saying the same thing is to state that the mass of uranium is greater than the combined masses of the fission fragments and that the excess mass is released as kinetic and radiated energy according to $E = mc^2$.

A nuclear reactor can produce a vast amount of useful electrical energy from a small amount of uranium fuel and a nuclear bomb (erroneously called "atomic") can produce an equally vast amount of destruction from the same amount of fuel. Both are **chain-reacting systems**. In such systems the crucial fact is that two or three neutrons are released with the fission fragments in each fission; these neutrons can then induce fissioning in other nuclei and the reaction becomes self-sustaining.

The energies involved when a sufficiently large number of uranium nuclei can fission are enormous and the destructive potential of nuclear energy soon became painfully obvious. All of this research was being conducted at the onset of World War II and a growing number of scientists, among them Albert Einstein, recognized that the allies were uncomfortably far behind Germany in harnessing this energy source. Scientists in the United States petitioned President Roosevelt for governmental aid to further research in this new field and investigation into the problem of controlling and harnessing fission was spearheaded by groups at the University of Chicago and the University of California at Berkeley. On December 2, 1942, the Chicago group, under the leadership of Enrico Fermi, made final preparations for an attempt at the first controlled, self-sustaining chain reaction. Late that evening in Chicago Arthur Compton phoned James Conant at Harvard and spoke in an impromptu code:

"The Italian navigator has landed in the New World."

"How were the natives?" asked Conant.

"Very friendly," replied Compton.

The age of nuclear energy had begun.

The *fusion reaction,* which is the basis for controlled thermonuclear reactions, is a method of obtaining enormous amounts of energy from hydrogen (like the hydrogen in sea water). It is also the reaction used in the hydrogen bomb and, further, the main reaction in our sun and stars that provides the light and heat to our planet. By a series of several nuclear reactions, four hydrogen nuclei or protons combine to form the nucleus of helium and, as we have seen, in doing so they release 28 MeV. So *fusion* is merely the nuclear "burning" of hydrogen to form helium, with a large release of energy. Thus far, attempts to use controlled thermo-nuclear reactions for peaceful means have not been success-ful, although a great amount of effort is being expended in this direction.

19-5

The Nuclear Force

As we already saw in Section 19-3, much about the nuclear force has been learned from working with the deuteron, although most of the information obtained to date is formidable and confusing. Remember that the Coulomb force depends only on the charges of the two particles involved and on the inverse square of their separation distance. It is believed that the nuclear force is not dependent upon particle charges, but its simplicity ends there. As Richard Feynman stated, "The nuclear interaction is as compli-cated as possible." It does not have any simple $1/r^2$ dependence on the separation distance, but is dependent instead upon the relative orientation of the spins, on the momenta of the particles, and on virtually every other known characteristic of the inter-acting nucleons. The gravitational force $1/r^2$ and the electrical force $1/r^2$ were simple. The magnetic force was merely a rela-tivistic correction of Coulomb's law. Until the nuclear force was discovered, these were the only known basic forces in nature. All other so-called forces, like those found in chemistry for the inter-action of molecules, could be derived from the basic electrostatic force—Coulomb's law. So nature had seemed simple; a $1/r^2$ law had explained everything. But the complexity of the nuclear force law is now causing many physicists to wonder if it is somehow inappropriate to utilize the force concept in describing the inter-action of nucleons. The nuclear force is so very complicated that it seems to go against the principle established by Copernicus and followed in all subsequent revolutions in physics—that nature is somehow basically simple. If our description (like the force be-tween nucleons) becomes too complicated, we may be describing nature incorrectly. We do not know whether this is true or not at this time; undoubtedly the problem in closely related to the dilemma we are presently facing in understanding elementary particles (discussed in the next chapter), since the two nucleons, the neutron and proton, are two of the many elementary particles currently under investigation.

19-6

Models of the Nucleus

As we have already mentioned, the difficulties of solving any quantum mechanical problem involving more than three particles are enormous. Even the largest computers available today cannot handle such problems. This is just as true of nuclear physics as it is of atomic, molecular, solid state, and other branches of applied physics. However, models can be constructed which will approximate the specific behavior determined by a particular kind of experiment. The construction of most models in nuclear physics thus far has been based on the mechanistic approach of Newtonian physics. Unfortunately, none of these models work well for all nuclei, but some do succeed in explaining the properties of restricted classes of nuclei for restricted types of experiments.

One, the "**Liquid Drop Model**" proposed by Niels Bohr and John Wheeler in the late thirties, works best in "explaining" nuclear fission for heavy nuclei. In this model, the nucleus is considered to be a spherical liquid drop with a more-or-less uniform charge density. If such a drop is disturbed (as when a uranium nucleus absorbs a neutron), the drop undergoes a violent jiggling, deforms severely into an elongated ellipsoidal shape, and finally breaks into two pieces of roughly equal size—the fission process (again, see Figure 19-3).

Another, the "**Shell Model**," arranges the nucleons in the nucleus in a shell-like structure somewhat similar to the shell structure of the electrons in an atom. A successful nuclear shell model was sought for many years, but was not found until 1949. Maria Mayer and Hans Jensen were eventually awarded the Nobel Prize in 1962 for the discovery.

The "Shell Model" can explain the large binding energies and therefore the great stability of certain nuclei like helium. In a manner analogous to the atomic shells, nuclear shells fill with particles which obey the Pauli exclusion principle and which have spin $\pm\frac{1}{2}$ (two spin orientations). But unlike the atomic analogue, the nucleus has two kinds of particles with these properties— neutrons and protons—which are not independent, but which interact with each other.

The simplest nucleus is that of hydrogen, given the symbol ^1_1H. In this notation the "1" on the lower left designates the positive charge—the total number of protons—while the number on the upper left gives the total number of nucleons, neutrons and protons, in the nucleus. The next stable nucleus is that of heavy hydrogen, the deuteron ^2_1H, which consists of one proton and one neutron. So far, the spin orientation of the particles appears to be unimportant since there is only one particle of each kind in the deuteron. (But this is misleading if we think of the nucleon as *one* kind of particle. The neutron-type and the proton-type are two possible "orientations" of the nucleon in some kind of "charge-space," similar to the two orientations of spin angular momentum in three-dimensional space. In fact, this seems to be true in nature.) It has been experimentally established that the spins of the two nucleons in the deuteron are aligned in the same

direction (Figure 19-4). If the spins were aligned in opposite directions, we would merely have a separate neutron and a separate proton because the nuclear force, which is dependent upon spin orientations, would not be strong enough to produce even one stable bound state.

The next heaviest nucleus is obtained by adding either a neutron or a proton to the deuteron. Adding a neutron gives us $_1^3$H, the nucleus of extra-heavy hydrogen or **tritium**. Adding a proton yields the nucleus $_2^3$He of the lighter isotope of helium. The spins of the two neutrons in $_1^3$H or the two protons in $_2^3$He must be in opposite directions (Figure 19-4) to satisfy the Pauli exclusion principle. By adding one more nucleon, we arrive at the very stable nucleus $_2^4$He which has two protons and two neutrons, each pair spinning in opposite directions (again, see Figure 19-4). This nucleus closes the first shell of proton states as well as the first shell of neutron states. (Note that $_2^3$He closed the first shell of proton states but not the first shell of neutron states and, conversely, that $_1^3$H closed the first shell of neutron states but not the first shell of proton states.)

As more neutrons and protons are added to build heavier nuclei, we find similar shell properties. Sometimes the proton shells are closed; other times the neutron shells are closed. In a few cases both the neutron and the proton shells are closed [as they were in $_2^4$He, (Figure 19-4)], producing particularly stable nuclei. Thus the "Shell Model" gives us a great deal of insight into the systematics of nuclear energies.

Unfortunately, it does not provide very much information about the other characteristics of the nuclei (the vibrations and oscillations that they can undergo, for example). It is also not very helpful in explaining how protons or other projectiles scatter around and through a nucleus during a collision process. In attempting to understand these other processes, physicists must rely on a number of nuclear models, some with strange and almost unbelievable properties. We seem to be at a sort of "patching" stage in our knowledge of the nucleus where each different class of nuclei has its own model. The modelmakers are continually trying to create a more general construction by combining the "good" features of some of the more restrictive models, but some physicists already believe that no one mechanistic model for the nucleus will ever be found. Just whose judgment is correct remains to be seen.

In any event, it appears certain that mechanistic models do indeed fail to explain the "structure" of the nucleons—the neutrons and protons that comprise the nucleus. Particle physics, one of the current fields of research, attempts to explain this particle structure.

Questions

1 Why was Rutherford's famous scattering experiment equally important to nuclear and to atomic physics?

2 Why does the existence of a very small, dense, positively-charged nucleus in the atom indicate that there might be a third force in nature (in addition to the forces of gravitation and electricity with which you are already familiar)?

3 Define an isotope.

4 What do physicists mean by a "nucleon"?

5 What is nuclear fission? Where is the energy which is released in fission generated?

6 What is nuclear fusion? Where is the energy which is released in this process generated?

7 Explain the importance of the discovery of the neutron.

8 The processes of both nuclear fusion and nuclear fission release energy. How is this possible?

9 Since the development of the atomic and hydrogen bombs, there has been considerable debate about a scientist's responsibility for the eventual applications of his work. To what degree do you feel that a scientist (or any other person) should be held responsible for the outcome of his ideas and work? Can a man be held responsible for the deeds of those beyond his control who use his ideas? Can he be held *not* responsible?

10 Explain how Einstein's mass-energy equivalence $E = mc^2$ is used to understand nuclear stability.

11 What important role does the Pauli exclusion principle play in governing nuclei?

Questions Requiring Outside Reading

1 Rutherford's experiments were important to both nuclear and atomic physics. Discuss some of his discoveries in the field of nuclear physics.

2 What were some of the earliest known phenomena of nuclear physics and why were they important?

3 What phenomena are referred to as "radioactivity"?

4 Otto Hahn, a German scientist, discovered nuclear fission during the 1930's. Read Hahn's view of the relationship between politics (or

government) and science and cite your reasons for agreeing or disagreeing with him.

5 Discuss the similarities and differences between the nuclear "Shell Model" and its atomic analogue.

6 Can you think of a reason why the neutron in a bound deuteron does not decay into a proton? Explain.

Problems

1 The mass of the tritium nucleus ^3_1H is about 5.009×10^{-24} g. Use the masses of the proton and neutron given in Example 19-1 (page 258) to find the binding energy of ^3_1H.

2 The nucleus of ^3_2He is the same as the ^3_1H nucleus except that a proton has been substituted for one of the neutrons. Since there are now two protons in the ^3_2He nucleus, would you expect its binding energy to be larger than, the same as, or less than that of ^3_1H? Explain.

3 What energy is released when a deuteron and a neutron form the nucleus of ^3_1H?

4 What energy is released when two deuterons combine to form ^4_2He? (*Hint:* The mass of ^4_2He is 6.644 g.)

5 Why do you suppose the mass of the neutron is slightly larger than the mass of the proton?

6 Five nucleons (of any combination and of either type) will not combine to form any nucleus. Can you give a reason for this? (*Hint:* Combine the fact that the nucleus forms a shell structure with the Pauli principle and the very short range of the nucleon force.)

7 A free neutron decays into a proton and an electron in an average time of about 12 minutes. If the electron's mass is about 10^{-27} g, what is the amount of kinetic energy released (in ergs and in MeV) in this decay process?

20

Elementary Particles

Introduction

In the past chapters, we have seen man's description of his physical universe evolve over 3000 years. First Aristotelian physics was replaced by the "New Physics" of the seventeenth century. Then the "Classical Physics" of Copernicus, Galileo, Newton, Faraday, and Maxwell was reformulated in the twentieth century by Planck, Einstein, Bohr, de Broglie, Schrödinger, Heisenberg, and others. If the modern revolutions in relativity and quantum mechanics could explain the results of every existing experiment today, we could tie all of physics up into neat little packages and say that we now "understand" the physical world around us. But the historical perspective shows us that this has never been possible before. In the past, whenever physicists thought that every discovery had been adequately explained and understood, some new experiment would crop up that simply did not fit the existing models and could not be explained by the old laws. The "clouds on the horizon" typify the state of physics today just as much as they did when Lord Kelvin alluded to them at the end of the nineteenth century—perhaps even more so because now a greater number of physicists are aware that the clouds exist.

A large amount of experimental information about elementary particles cannot be understood or organized by any of the previous "laws" we have learned. We are apparently in the midst of a new revolution in physics. It would be nice to be able to predict its outcome, but obviously we cannot anticipate what new ideas will be generated or what changes in currently accepted thought will occur in even the near future. We probably do not yet know the correct language in which the ideas of the new revolution will

be expressed, but we can outline the novel experimental facts that already exist and discuss a few of the empirical laws or explanations that have recently been proposed to help organize these still inexplicable facts.

Particle physics came into existence in the late 1890's with the discovery of the electron by J. J. Thompson and the discovery of the photon—the quantum or particle of electromagnetic radiation—by Planck and Einstein. By the mid-1920's there were three known elementary particles—the **electron**, the **photon**, and the **proton** (which we already know is the nucleus of the hydrogen atom, as well as a main constituent of heavier nuclei). The first piece of evidence that particle physics was not as simple as this was the discovery by Uhlenbeck and Goudsmit in 1925 that two "kinds" of electrons were required to explain the electron structure in atoms. As we saw in Section 18-7, page 250, electrons can orient themselves either "up" or "down" relative to some fixed direction; further, each electron acts like a tiny, submicroscopic magnet—just as if it were a charged sphere spinning on its axis. Although it was still not clear why electrons possessed these unusual characteristics, Uhlenbeck and Goudsmit's electron spin theory was needed to explain the atomic shell structure of electrons, the arrangement of the periodic table, and the shell model of the nucleus.

20-2

When Physics Was Simple— Only Three Elementary Particles

The first real step toward understanding came in 1928 when a 26-year-old English physicist, P. A. M. Dirac, proposed a theory that not only explained electron spin, but unexpectedly led to a novel second prediction which confounded the entire world of physics. In Dirac's words:

20-3

Dirac and Antiparticles

> The new quantum mechanics, when applied to the problem of the structure of the atom with point-charge electrons, does not give results in agreement with experiment. The discrepancies consist of "duplexity" phenomena, the observed number of stationary states for an electron in an atom being twice the number given by the theory. To meet the difficulty, Goudsmit and Uhlenbeck have introduced the idea of an electron with a spin angular momentum. This model for the electron has been fitted into the new mechanics by Pauli.
>
> The question remains as to why Nature should have chosen this particular model for the electron instead of being satisfied with the point-charge. One would like to find some incompleteness in the previous methods of applying quantum mechanics to the point-

DIRAC P. A. M. (Paul Adrien Maurice) Dirac was born in 1902 in Bristol, England. He received his secondary education there and obtained a degree in electrical engineering from the University of Bristol in 1921. Dirac then decided to change his field of study to mathematics and, to do so, remained at Bristol for two more years before going to Cambridge to work on his Ph.D. □ Dirac received his doctoral degree in 1926 and continued to work in the new physics which had arisen in Germany one year before. His theory of electrons and positrons, published in 1928, earned him many honors. In 1932, Dirac was appointed the Lucasian Professor of Mathematics at Cambridge, the chair Newton had once occupied, and a year later won the Nobel Prize (with Schrödinger) for his electron-positron theory. □ Like Schrödinger, Dirac liked to work alone, but unlike Schrödinger, Dirac was very mathematical and abstract. (He once told Schrödinger to beware of models and pictures in quantum mechanics.) Dirac's personal life was also well ordered and precise. He believed that a sentence should not be begun until it had an end, and practiced what he preached. He used language so sparingly it is reported that he sometimes believed even a simple "yes" or "no" to be excessive. □ Today Dirac is recognized as one of the great men in the development of modern physics, and his travels and lectures often take him away from England to other countries, including the United States. He is currently affiliated with Florida State University.

P. A. M. Dirac

The Niels Bohr Library, American Institute of Physics

charge electron such that, when removed, the whole of the duplexity phenomena follow without arbitrary assumptions. In the present paper it is shown that this is the case, the incompleteness of the previous theories lying in their disagreement with relativity. It appears that a point-charge electron satisfying the requirements of relativity leads to an explanation of all duplexity phenomena without further assumption.[1]

Dirac noted that the relation between energy and momentum in Einstein's relativity is a quadratic relation; that is, it is a mathematical relation between the square of the energy E and the square of the momentum p (see Chapter 16, page 204):

$$E^2 = p^2c^2 + m^2c^4 \tag{20-1}$$

Yet Dirac also knew that the basic mathematical structure of quantum mechanics required that the energy of a particle be expressed in terms of the *first power* of E. Thus, there appeared to be a basic incompatibility between relativity and quantum mechanics. If the square root of the energy-momentum equation above is written, the first power of E is obtained, but only at the expense of a square root on the right-hand side of the equation

$$E = \pm \sqrt{p^2c^2 + m^2c^4} \tag{20-2}$$

[1]"The Quantum Theory of the Electron," *Proceedings of the Royal Society,* A117 (London, 1928): p. 610.

which is difficult to interpret physically. In the very structure of the rules of quantum mechanics, such a square root simply did not make any sense. According to quantum mechanics, a linear or first power equation involving the energy was needed. But how could such an equation be relativistic [i.e., be compatible with Equation (20-1)]?

Dirac proposed an equation with the linear form

$$E = Apc + Bmc^2 \qquad (20\text{-}3)$$

where A and B were as yet unspecified constants. Now it is obvious from elementary algebra that if A and B are ordinary constants (numbers), this equation will not be compatible with the energy-momentum relation from relativity, Equation (20-1). If we square both sides of Equation (20-3),

$$E^2 = A^2p^2c^2 + B^2m^2c^4 + 2ABpcmc^2 \qquad (20\text{-}4)$$

we do not arrive at Equation (20-1) as required by relativity because of the cross-product term $2ABpcmc^2$. However, as Dirac showed, if A and B were not ordinary constants or numbers in the usual sense, then Equation (20-3), when squared, could be *forced* to be equal to the correct relation of Equation (20-1) as required by special relativity. The result was that A and B each turned out to be what mathematicians call a **matrix**.[2] In this case, *four* distinct wave functions ψ must be associated with A and B. Hence, Dirac had to conclude that there were *four* "kinds" of whatever particle his Equation (20-3) described, two corresponding to the two allowed "up" and "down" spin orientations of the electron and two corresponding to two similar orientations of a new particle—just like the electron but with an *opposite charge!* In this way, Dirac predicted the existence of the **positron**—the **antiparticle** of the electron. (The positron itself was not discovered experimentally until 1932.) We quote from Dirac's explanation of his work:

> The question that we must first consider is how theory can give any information at all about the properties of elementary particles. There exists at the present time a general quantum mechanics which can be used to describe the motion of any kind of particle, no matter what its properties are. The general quantum mechanics, however, is valid only when the particles have small velocities and fails for velocities comparable with the velocity of light, when effects of

[2]A matrix is the mathematician's extension of the concept of a vector. A matrix is related to a vector in much the same way that a vector is related to a scalar quantity (see page 42). In two dimensions, for example, a scalar quantity is defined by a single number, while a vector requires two numbers (length and direction). Then, in this particular instance, a matrix would require 2×2 or 4 numbers to define it. For Dirac's case, the vector has four components and the corresponding matrix has 4×4 or 16 numbers.

relativity come in. There exists no relativistic quantum mechanics (that is, one valid for large velocities) which can be applied to particles with arbitrary properties. Thus when one subjects quantum mechanics to relativistic requirements, one imposes restrictions on the properties of the particle. In this way one can deduce information about the particles from purely theoretical considerations, based on general physical principles.

This procedure is successful in the case of electrons and positrons. It is to be hoped that in the future some such procedure will be found for the case of the other particles. I should like here to outline the method for electrons and positrons, showing how one can deduce the spin properties of the electron, and then how one can infer the existence of positrons with similar spin properties and with the possibility of being annihilated in collisions with electrons.[3]

The positron—the antiparticle of the electron—is positively charged. It has the same quantity of charge and the same mass as the electron. The novel characteristic the positron exhibits experimentally is that it can annihilate mutually with an electron so that the total energy (mass energy plus kinetic energy) of each particle is transformed into **gamma rays**—high energy photons of electromagnetic radiation. The total energy of these gamma photons is equal to the total mass (plus kinetic energy) of the original electron and positron:

$$e^+ + e^- \rightarrow \gamma \text{ photons}$$

Thus, the electron-positron theory is linked with the quantum of electromagnetic radiation—the photon.

Forcing the relativity equations to fit the mathematics of quantum mechanics led Dirac to an entirely new concept of the composition of the universe. In other words, for an electron to be described in relativistic language and retain its quantum mechanical wave properties, that electron *must* have a spin of $\frac{1}{2}$ and it *must* have an antiparticle.

In his Nobel Prize address in 1933, Dirac speculated on the cosmological significance of the symmetry between electrons and positrons:

If we accept the view of complete symmetry between positive and negative electric charges so far as concerns the fundamental laws of Nature, we must regard it rather as an accident that the Earth (and presumably the whole solar system) contains a preponderance of negative electrons and positive protons. It is quite possible that for some of the stars it is the other way about, these stars being built up mainly of positrons and negative protons. In fact, there may be half the stars of each kind. The two kinds of stars would both show exactly the same spectra, and there would be no way of distinguishing them by present astronomical methods.[3]

[3]P. A. M. Dirac, "Theory of Electrons and Positrons," *Nobel Lectures* (New York: American Elsevier Publishing Co.), p. 320. Copyright © The Nobel Foundation, 1934.

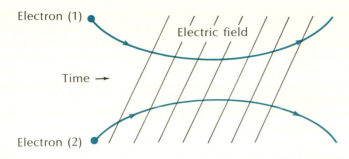

20-1 Two charged particles scattering in the classical Coulomb field.

With the success of Dirac's theory, many physicists began searching for a more complete theory that would encompass electrons, positrons, *and photons*. One was finally obtained in the late 1940's by several men independently, among them Richard Feynman who now teaches at the California Institute of Technology. This theory, given the formidable name of quantum electrodynamics (abbreviated QED) provides a semimechanistic "picture" of electromagnetism—a description of the electromagnetic field in which Faraday and Maxwell's ether is not required.

In quantum electrodynamics, photons of electromagnetic radiation are emitted by accelerated electrons and by electrons which "move" from a higher energy level in an atom to a lower one. These photons are real in the sense that they can be absorbed and registered by an observer (either by sight or by using some other medium such as photographic film). In addition, QED also provides a picture of Newton's "action-at-a-distance" and of Maxwell's electric field in terms of **virtual** photons which are not visible. Figure 20-1 represents the old Maxwell-Faraday picture of the electromagnetic field; in it electron (1) is repelled by electron (2) due to their mutual electric field, resulting in the trajectories shown. Feynman and others replaced the ether—which they no longer considered to be a viable construct—with electromagnetic photons, the quanta of the electric field. These virtual photons serve only as agents to transmit the force between particles. In Figure 20-2 (called a "Feynman diagram"), for example, electron (1) emits a photon (represented by the dashed line) as it moves

20-4

The Modern Theory— Quantum Electrodynamics

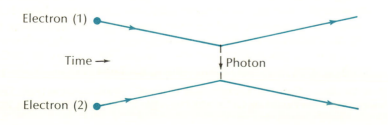

20-2 Feynman diagram showing two charged particles scattering in a quantum electrodynamic (QED) field.

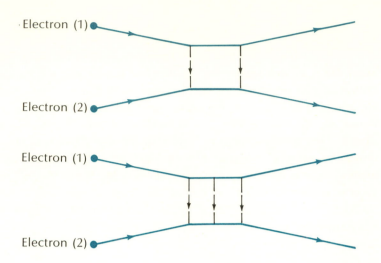

Electron (1)

Electron (2)

Electron (1)

Electron (2)

20-3 Higher-order Feynman diagrams showing the emission and absorption of two and three photons.

along. Since the photon has momentum, electron (1) must recoil after the emission. The photon is then absorbed by electron (2), which must also recoil when it *gains* the momentum of the photon. So we see that the two electrons repel one another, not because of an electric field in an all-pervading ether, as formerly believed, but because the electromagnetic photon acts as an agent in transferring the change in momentum from electron (1) to electron (2).

In addition to this process, two, three, or more photons could be emitted (see the higher-order Feynman diagrams in Figure 20-3) or a single electron could emit a photon and then *reabsorb* it (Figure 20-4). If one of the electrons were replaced with a positron,

20-4 Virtual emission and reabsorption of a photon by one electron.

Photon

Electron

an attractive rather than a repulsive force would exist. Many, many experimental tests of this rather attractive and simple theory have been made and, to date, not one experiment has clearly violated the rules of quantum electrodynamics. So we accept QED as the end product of the Dirac theory and as a successful theory of electrons, positrons, and photons.

20-5

More Elementary Particles

Unfortunately, no successful theory has yet been compiled which explains the remainder of the elementary particles and their interactions with one another. Three more particles—the neutrino, the neutron, and the muon—were discovered in the 1930's and one new particle—the **pi-meson** or **pion** π—in the 1940's. The pion is of particular interest since its properties were predicted by a

FEYNMAN

Richard Feynman is one of the most colorful physicists of the present generation. He is as well known for his sense of humor and his practical jokes as he is for his contagious enthusiasm for physics. Feynman was born in 1918 in New York City and attended public schools there. He attended M.I.T. where he earned his Bachelor of Science and then went on to earn his Ph.D. at Princeton. ☐ After his Ph.D. was awarded in 1942, Feynman joined the "Manhattan Project" at Los Alamos, New Mexico where he helped in the development of the atomic bomb. ☐ Although he was only in his twenties, Feynman was a group leader at Los Alamos and was as well known for his practical jokes as he was for the effectiveness of his work. From time to time, for instance, it was reported that security guards on routine checks of the filing cabinets would find that someone had picked one of the locks, left some cryptic message, and locked the cabinet again. In this way he was protesting against the insufficient security measures which left such important secrets in files so easy to open. ☐ In the past few years Feynman's enthusiastic lecturing style has been captured in the form of a series of motion pictures used in many college physics courses. In addition, he has taught the basic physics course at Cal Tech and the published transcripts of his lectures, *The Feynman Lectures in Physics,* have been used by physics students around the world. ☐ In 1965 Feynman, Schwinger of Harvard, and Tomanaga of Japan were awarded the Nobel Prize for their independent contributions to the development of quantum mechanics.

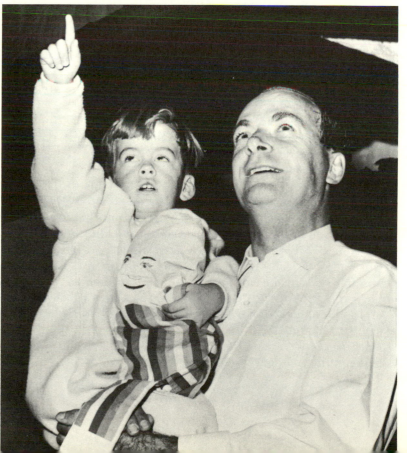

Wide World Photos

Nobel prize winner Richard Feynman with his son, Carl, October 21, 1965.

Japanese physicist, H. Yukawa, in 1936 before it was experimentally discovered. Yukawa reasoned that the pion was needed to provide the strong nuclear force between two nucleons—a neutron and a proton; that is, that the two nucleons interact by exchanging a pion in much the same way two electrons interact through the Coulomb force by exchanging a photon. But the pion differs from the photon in that it has a definite nonzero rest mass (a photon has a zero rest mass). Consequently, Yukawa theorized that the nucleon-nucleon force must be much shorter in range than the Coulomb force. Thus, the pion already had a theoretical reason for existing, even before its discovery—a statement which we will find ourselves unable to make about most of the later particles still to be discovered at this point.

In the 1950's a spectacular and unexpected increase in the number of elementary particles occurred, primarily because of the building and operation of high-energy particle accelerators such as the Bevatron at Berkeley. Protons could now be accelerated to extremely high energies (billions of electron volts or BeV). Aiming a beam of high-energy protons at other target protons, usually in the form of liquid hydrogen, and other energetic encounters, resulted in many "strange" new particles. By 1960 the list of elementary particles had grown to about 30. Some of these were antiparticles such as the antiproton and the antineutron, but most of them had at least somewhat novel properties. For instance, *any number* of pi-mesons or pions can be produced or created from a proton-proton collision. The number of pions is limited only by the amount of kinetic energy involved in the collision, since the creation of the rest mass of each pion requires an initial proton kinetic energy at least equal to the rest mass of the created pion times the square of the velocity of light ($E = mc^2$).

On the other hand, protons and neutrons (like electrons) cannot be created at will. They can be created from energy only if their corresponding antiparticle appears at the same time. A fundamental difference seems present here. Scientists have given the name **mesons** to particles like the pion, **baryons** to particles like the proton and neutron, and **leptons** to the electron and other light particles—the neutrino and muon. Experiments have revealed that the number of baryons is conserved; it will be the same in any collision, while the number of mesons need not be. These rules work automatically if the proton and neutron are given a **baryon number** of +1, the antiproton and the antineutron are given a baryon number of −1, and the meson is given a baryon number of 0. (Note the similarity here between baryon numbers and electric charges. With these rules, the total baryon number must be conserved—remain the same—in any reaction, just as the total charge must be conserved in a reaction.)

By the mid-1950's, more and more data concerning new elementary particles had accumulated. A new meson heavier than the pion, the K-meson, and several new baryons heavier than the neutron or proton—the lambda (Λ), sigma (Σ), and xi (Ξ) particles—had all been discovered. The new particles were not stable like the proton, but decayed radioactively in a fraction of a second. They had been produced by bombarding protons with high-energy pions. A typical reaction is one in which a K-meson and a Σ-baryon are produced when a π-meson bombards a proton:

$$\pi^+ + p^+ \rightarrow K^+ + \Sigma^+$$

(The plus signs indicate that the particles are positively charged.)

A confusing state of affairs arose when experimentalists found that some of these reactions were simply never seen. For instance, a π^--meson bombarding a proton would not yield a Σ^+ and a π^-, although the expected reaction ($\pi^- + p^+ \rightarrow \Sigma^+ + \pi^-$) did not violate any known **conservation laws**. Conservation of energy, momentum, angular momentum or spin, charge, and baryon number all held—yet the reaction was never seen. It seemed to experimentalists that, in such a reaction, some kind of hidden "charge" which is not conserved must be carried by the particles.

Additional experiments only served to make the matter even more confusing. The Λ and Σ particles would decay into a neutron or a proton by emitting a π-meson:

$$\Lambda^0 \rightarrow p^+ + \pi^-$$

or

$$\Lambda^0 \rightarrow n^0 + \pi^0$$

or

$$\Sigma^+ \rightarrow p^+ + \pi^0$$

and so on. However, all of the particles involved here (p, n, Λ, Σ, and π) are what are called **strongly-interacting particles** or **hadrons** because of the strength of their nuclear force. A typical energetic hadron interaction should be completed in about the time that light, with its velocity of $c = 3 \times 10^{10}$ cm/sec, takes to traverse the particle's diameter d—about 10^{-13} cm. This typical time turns out to be

$$t = \frac{d}{c} = \frac{10^{-13} \text{ cm}}{3 \times 10^{10} \text{ cm/sec}} = 0.33 \times 10^{-23} \text{ sec}$$

or about 10^{-23} sec in round numbers. But all experiments in decays involving a π-meson have shown typical times which were much, much longer—about 10^{-10} sec. (This may seem to be an extremely short time to us, but it is actually a very long time by nuclear

Wide World Photos

Murray Gell-Mann with his wife, Margaret, and son, Nicholas, after learning he had won the Nobel Prize, October 31, 1969 (Halloween).

standards.) Likewise the Ξ decayed into a Σ or a Λ by emitting a pion in the same characteristically long time of 10^{-10} sec. Something seemed to be preventing the change of a Σ or a Λ into a p or an n and delaying the change of a Ξ into a Σ or a Λ. What was nature trying to tell us this time?

Murray Gell-Mann, who also teaches at Cal Tech with Feynman, worked on these two apparently unrelated problems of the absence of certain reactions and the "delayed" π-meson decay until he found a solution in 1954. Gell-Mann reasoned that if the Σ and Λ particles had a new and as yet undiscovered quantum number which differed from that for the proton, neutron, and pion, then the inhibition of the reaction

$$\pi^- + p^+ \rightarrow \Sigma^+ + \pi^-$$

could be explained by the fact that the reaction violates the conservation of this new number. The "slow" decay of the Λ particle into $p^+ + \pi^-$ could also be understood for the same reason. Since the decay takes so much longer (10^{-10} sec) than expected (10^{-23} sec), it can be said that the decay reaction is delayed by a factor of $10^{-10}/10^{-23}$ or 10^{13}. This is such an extremely large number that we can say that the Λ particle effectively decays into

GELL-MANN

Murray Gell-Mann has been described by a biographer as one of the few living successors to the mantle of Albert Einstein. Gell-Mann's contributions to particle physics have been extraordinarily extensive, imaginative, and creative. ▫ Gell-Mann was born in 1929 in New York City and attended public school there until the age of eight when he was transferred to a private school. Throughout, he considered the classroom material "dull, dull stuff"—an observation, incidentally, that Einstein had made about his own schooling half a century earlier. ▫ Gell-Mann went through the standard academic preparatory program for college and entered Yale University at the tender age of 15. Professors there described him as "sleepy, perhaps bored." According to Gell-Mann, his young age in comparison with his classmates "was a disadvantage, since my personality was not yet well developed." After graduating, he took the usual steps up the academic "ladder" in an unusually short time since his contributions to physics were already beginning to be recognized on a worldwide basis. He finally accepted the offer of a full professorship at the California Institute of Technology in 1956 at the age of 26. It is said that, among other reasons, Cal Tech wanted Gell-Mann to teach there so that Feynman would have someone around to talk to. ▫ Gell-Mann is best known for his invention of the "strangeness" quantum number and for his organization of elementary particles via the mathematics of group theory—the "Eight-Fold Way," as Gell-Mann called it. He was awarded the Nobel Prize in 1969 for both contributions. However, Gell-Mann's interest in physics is very broad, as is his knowledge in other fields, both cultural and scientific. ▫ He is widely read, skis, hikes, and is an avid birdwatcher. He has versatile tastes in food, and the author remembers dining with Gell-Mann once in San Francisco when he ordered an entire Chinese meal—in Chinese (a somewhat spectacular feat for a physicist). ▫ Gell-Mann's contributions to physics will most likely be remembered for a long, long time to come. His "Eight-Fold Way" organization of particles has proved to be such a significant breakthrough in finding empirical laws that is has been compared with Balmer's formula for the hydrogen spectrum, Mendeleev's periodic table, and Kepler's laws for planetary motion.

p + π only very rarely, since the Λ has a new quantum number which differs from that for the p + π combination. Note that for some strange reason nature does not absolutely forbid the decay of Λ into p + π; she merely slows it down by a factor of 1 to 10^{13}.

Gell-Mann called this new quantum number for the Σ and Λ particles **strangeness**. We will now attempt to define strangeness. The energy-level diagram shown in Figure 20-5, plots the rest masses (energies) of the baryons known to Gell-Mann at the time. How are the nucleons n and p different from the Λ particle or the Σ family Σ^+, Σ^0, Σ^-? The nucleons have two possible charges, 0 and +1; the Σ's have three, −1, 0, and +1, and the Λ has only one, 0. The Σ family and the Λ have one characteristic in common: The center of charge (the average charge) of the Σ family (0) is the same as the charge of the Λ, but the nucleon (n-and-p) center of charge is $+\frac{1}{2}$. Gell-Mann reasoned that this property—the center

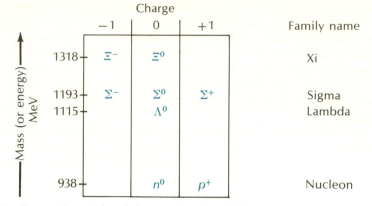

20-5 Energy (mass)-level diagram for the eight baryons.

of charge of the Σ **family** or **multiplet**—was responsible for the strange behavior in particle interactions. Past experience in quantum mechanics has shown that quantum numbers change only by integer values. To make strangeness an integer, the center of charge must be multiplied by 2 so that the strangeness changes by 1 as it travels from the Λ particle to the nucleon. We also subtract 1 to give the nucleons strangeness $S = 0$ (for historical reasons). The Λ and the Σ particles then have $S = -1$ and the Ξ particle multiplet, with center of charge at $-\frac{1}{2}$, has $S = -2$.

The mesons are classified in a similar way (see Figure 20-6). The pion family has three members—π^+, π^0, π^-—with a center of charge at zero and strangeness defined as $S = 0$. (We do not subtract 1 in the case of mesons, again for historical reasons.) There are two distinct families of K-mesons. The K^- and $\bar{K}^0$ have their center of charge at $-\frac{1}{2}$ and a strangeness of -1. The K^0 and K^+ have their center of charge at $+\frac{1}{2}$ and a strangeness of $+1$. We see that, for the mesons, strangeness is simply defined as twice the value of the center of charge.

Now we can use these empirical rules to find out why the

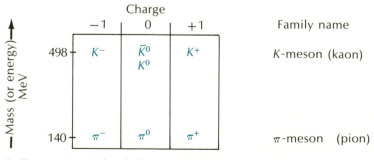

20-6 Energy (mass)-level diagram for the seven mesons.

reaction $\pi^- + p^+ \rightarrow \Sigma^+ + \pi^-$ will not occur. The total strangeness of the particles on the left-hand side of the reaction is 0, while the total strangeness of the particles on the right-hand side is -1 (π has $S = 0$, while Σ has $S = -1$). Therefore, strangeness is not conserved in the reaction. On the other hand, the reaction $\pi^+ + p^+ \rightarrow K^+ + \Sigma^+$ *can* occur since both sides of the reaction have $S = 0$ (on the right-hand side, K^+ has $S = +1$ and Σ^+ has $S = -1$, totalling zero).

But why should strangeness be a new quantum number that is conserved in hadron reactions as are charge, energy, momentum, baryon number, and spin? Why should the center of charge of a family determine this novel property of elementary particles? Physicists today cannot answer these questions any more than Kepler could have answered *why* planets travel in elliptical orbits or Balmer could have answered *why* the frequencies of light from hydrogen are dependent upon the square of an integer. However, the empirical rule of conservation of strangeness has held true in elementary particle physics for almost two decades and so we accept strangeness as a valid property of the particles that make up the world.

Since 1960 the number of elementary particles has grown even larger until it now totals over 200. The term "elementary" begins to lose its meaning for most people when it is applied to such a vast number of particles. Numerous attempts were made during the late 1950's and the 1960's to classify these particles and to predict their existence by means of simple formulas analogous to the Balmer formula which predicted the discrete color lines in the hydrogen spectrum. One of these attempts met with spectacular success. A method obtained from a branch of mathematics called "group theory" was devised independently by both Murray Gell-Mann in the United States and Yuval Ne'eman in Israel in the early 1960's. We will see in the next section that it is fairly easy to understand the Gell-Mann—Ne'eman scheme by using a semimechanistic model which does not require an understanding of the mathematics of group theory. Because of his work on this method and his earlier discovery of strangeness, Gell-Mann was awarded the Nobel Prize in 1969.

First let us return to the energy-level diagrams for the eight baryons. The arrangement of particle multiplets that nature presents us with is certainly not random or haphazard. There is a symmetry in Figure 20-7 that indicates some underlying order. This diagram represents the basic **octet** of particles (each with a spin of $\frac{1}{2}$) as predicted by Gell-Mann's "Eight-Fold Way" theory. Two families of **doublets**—the nucleon and the Ξ multiplets—

20-7

Still More Particles and the "Eight-Fold Way"

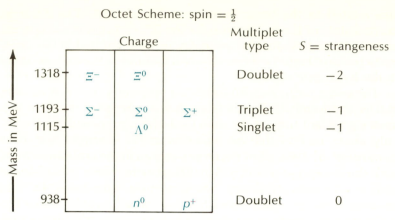

Octet Scheme: spin = $\frac{1}{2}$

	Charge			Multiplet type	S = strangeness
1318	Ξ^-	Ξ^0		Doublet	−2
1193	Σ^-	Σ^0	Σ^+	Triplet	−1
1115		Λ^0		Singlet	−1
938		n^0	p^+	Doublet	0

Mass in MeV

20-7 The octet group of baryons with spin = $\frac{1}{2}$—Gell-Mann's "Eight-Fold Way."

together exhibit a kind of symmetry around a zero charge. That is, the nucleon has two members of 0 and +1 charge, whereas the Ξ has two members of 0 and −1 charge. The Σ particles have three charge states of +1, 0, and −1 and are therefore symmetric about a zero charge, as is the singlet Λ particle. These eight particles exhibit a striking symmetry that almost cries out for an explanation—an explanation which came when Gell-Mann's theory predicted the existence of these eight particles.

As more particles were discovered in the early 1960's, Gell-Mann's theory also predicted that a group of these heavier baryons would fit into a **decuplet**—a scheme of ten particles, each of which has the same spin. The grouping of ten was a natural consequence of group-theory mathematics. The decuplet, as we know it today, appears in Figure 20-8. The asterisks in this figure represent parti-

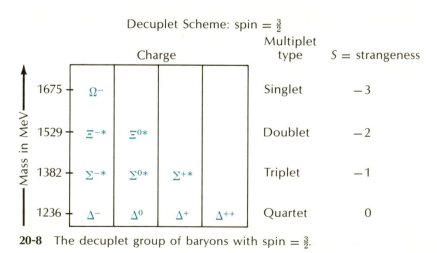

Decuplet Scheme: spin = $\frac{3}{2}$

	Charge				Multiplet type	S = strangeness
1675	Ω^-				Singlet	−3
1529	Ξ^{-*}	Ξ^{0*}			Doublet	−2
1382	Σ^{-*}	Σ^{0*}	Σ^{+*}		Triplet	−1
1236	Δ^-	Δ^0	Δ^+	Δ^{++}	Quartet	0

Mass in MeV

20-8 The decuplet group of baryons with spin = $\frac{3}{2}$.

cles related to the basic baryons in Figure 20-7. The Σ^* is an excited state of the Σ; likewise for the Ξ^*. Although basically the same with or without the asterisks, in their excited states these particles have higher energies (masses) similar to the excited states in the hydrogen atom.

These ten particles also exhibit a symmetry, but a symmetry that is apparently different from that in the octet scheme. It is most significant that, at the time Gell-Mann proposed his theory, only about half of the ten particles in the decuplet had been discovered. By 1964, all of the empty entries in the table were filled with the exception of the omega (Ω) particle. Since it is a singlet with a -1 center of charge, the Ω particle must have the very unusual value of strangeness $S = -3$, which would make it a difficult particle both to produce and to observe in reactions where protons—particles with $S = 0$—are used as targets. The Ω particle would also have a very characteristic decay pattern. For instance, it might decay by emitting a π-meson into an $S = -2$ particle which would, in turn, decay into an $S = -1$ particle and a π; finally the $S = -1$ particle would decay by emitting a pion into an $S = 0$ neutron or proton.

This was a very exciting time in high-energy physics. A feverish search for the Ω^- began at the two highest energy accelerators in the world—The Brookhaven Laboratory on Long Island and CERN in Geneva, Switzerland—which housed the only two machines of high enough energy to produce the Ω^-. Late in 1964, the Brookhaven group found the Ω^- particle; it possessed the exact charge, mass, and strangeness Gell-Mann had predicted. The Ω^- is such an unusual particle to find that, even by 1973, only several dozen Ω particles had been identified and "discovered."

20-8

The Quark

Gell-Mann took the name for still another new "particle" from the line "Three quarks for Muster Mark!" in James Joyce's *Finnegan's Wake*. It appears that Gell-Mann's contributions to physics are as exceptional as his knowledge of twentieth century novels and Zen Buddahism (his Eight-Fold Way). Gell-Mann proposed that every baryon is composed of three basic "fundamental" particles—the quarks. With this construct, we can build a simple model to illustrate the decuplet scheme. We will call these three quarks a, b, and c and assign them the following charges:

symbol of quark	a	b	c
charge of quark	$-\frac{1}{3}$	$-\frac{1}{3}$	$+\frac{2}{3}$

We will further assume that the mass of quark b equals the mass of c, but that the mass of quark a is heavier than b or c by 146 MeV. Now the ten possible combinations of these three basic "building blocks" for elementary particles are: (aaa), (aab), (abb), (bbb), (aac), (acc), (ccc), (bcc), (bbc), (abc). If these combinations are arranged according to their masses, (aaa) will appear alone as the heaviest—a singlet, the Ω^-, of charge -1 and therefore of strangeness $S = -3$. Next, (aab) with a charge of -1 and (aac) with a charge of 0 each have a mass 146 MeV lighter than (aaa). (aab) and (aac) correspond to the Ξ^* particles. Likewise, (abb), (abc), and (acc) give the Σ^* particles. Finally, (bbb), (bbc), (bcc), and (ccc) give the delta Δ quartet of particles. So the charges, masses, strangeness quantum numbers, and multiplet families of the decuplet (Figure 20-8) can be "understood" by this simple quark model. Although we will not do so here, the octet (Figure 20-7) can also be obtained from the three quarks, and the several meson groups can be obtained from a quark plus an antiquark. In fact, as of 1973 all baryon and meson multiplets could be explained by this model.

The quark model, with its success in predicting the groupings of hundreds of elementary particles, appears to be very sound. However, no *individual* quark has ever been observed, although a great deal of effort has been devoted to locating one. Quarks have fractional charges and so they should not be too hard to find if they appear in nature. Most physicists today have the general feeling that, although the quark scheme gives a good mechanistic picture of the group theory underlying the organization of the baryons and mesons, the quarks themselves are probably not "real" in the sense that elementary particles are actually composed of quarks. The abstract mathematics of the group theory that leads to the same result as the quark model is "real," but thus far the quark model is not understandable on any fundamental level other than that of a workable empirical scheme.

20-9

Symmetry

The symmetry principles or "conservation laws" of physics have become increasingly important in elementary particle research in the past two decades. Before 1956, a particular list of conservation laws was assumed to hold for all particle interactions on a microscopic level. Perhaps physicists turned to the conservation laws partly because the fundamental force laws (with the exceptions of Coulomb's law for electric charge forces and Newton's gravitational law for gravitational forces) between the elementary particles were just too incredibly complicated. The conservation laws until 1956 were:

Conservation of Energy
Conservation of Momentum
Conservation of Angular Momentum
Conservation of Spin Quantum Number

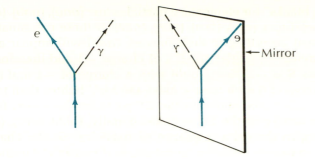

20-9 An electron emitting a photon and the mirror image of this process.

Conservation of Charge
Conservation of Strangeness
Conservation of Baryon Number
Conservation of Parity (P)
Conservation of Charge Conjugation (C)
Conservation of Time Reversal (T)

The last three laws may be considered symmetries. **Parity**, for instance, is merely the operation of describing an elementary particle reaction in terms of its mirror image (that is, in terms of how it would appear if observed in a mirror). If parity were conserved, as everyone thought it was, then the mirror image of any reaction would still be a possible physical process. This may seem like a trivial statement since it simply means that the physical laws of the universe are not intrinsically left-handed or right-handed. For instance, let us imagine the mirror image of the process of an electron emitting a photon (a γ-ray). The mirror image of this is certainly an acceptable physical process as well, since it is described by the same physical laws as the original process (see Figure 20-9). Now consider an electron spinning on its axis and moving with a velocity $\vec{v}$ along its spin-angular-momentum direction. What must be remembered here is that the mirror image of a spinning object is seen spinning in the opposite

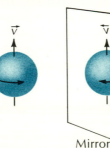

Counterclockwise
spin *up*
velocity *up*

Clockwise
spin *down*
velocity *up*

Mirror

20-10 An electron moving in the direction of its spin and the mirror image of this process.

direction. Therefore, according to the conservation of parity law, electrons spinning counterclockwise and moving in the same direction as the spin vector must be as equally probable as electrons spinning clockwise with a spin direction opposite to the velocity.

In 1956 two theoretical physicists, T. D. Lee and C. N. Yang, were trying to understand a problem in particle physics that could not be explained by the rules of conservation of parity. They suggested a crucial experiment, performed at Columbia University the same year, to test the parity "law." The experiment was concerned with detecting electrons emitted from radioactive nuclei in such a way that the spin of the electrons was oriented (by orienting the nuclear spin) in a fixed direction (say, counterclockwise or up). If parity were a valid principle, they reasoned, then just as many electrons would appear with a downward velocity below the nucleus as would appear with an upward velocity above the nucleus. But the experiment showed that fewer electrons were emitted downward than were emitted upward; hence Lee and Yang concluded that nature does not conserve parity in the so-called **weak interactions** such as nuclear radioactive beta decay (electron emission) and the decay of the strange particles by pion emission. Experiments were then conducted to test the conservation of parity law in strong interactions such as the bombardment of protons by pions, other protons, K-mesons, and so on; in these cases parity was found to be conserved. Lee and Yang received the Nobel Prize in 1957, within a year of the publication of their paper which stated these conclusions:

> The conservation of parity is usually accepted without questions concerning its possible limit of validity being asked. There is actually no a priori reason why its violation is undesirable. As is well known, its violation implies the existence of a right-left asymmetry.
>
> One may question whether the other conservation laws of physics could also be violated in the weak interactions. Upon examining this question, one finds that the conservations of the number of heavy particles, of electric charge, of energy, and of momentum all appear to be inviolate in the weak interactions. The same cannot be said of the conservation of angular momentum and of parity. Nor can it be said of the invariance under time reversal.[4]

Time reversal conservation has to do with reversing the direction of time in microscopic particle reactions. We run the movie camera backward, so to speak, to see if the same laws of physics describe the time-reversed situation. **Charge conjugation** is the replacement of every particle with its antiparticle (its opposite charge). Again, it was believed that these symmetry principles held

[4]T. D. Lee and C. N. Yang, "Question of Parity Conservation in Weak Interactions," *Physical Review*, 104 (1956): p. 258.

in the world, that the same "laws" of physics would describe the interaction between particles when the time direction was reversed and when all particles were exchanged for antiparticles.

A flurry of experimental activity followed Lee and Yang's discovery in 1956; physicists reasoned that since parity was not conserved in weak interactions, perhaps some of the other conservation "laws" might not hold under all conditions. Every conservation law was checked and, in 1964, a group of experimentalists at Princeton University found a second violation of the conservation laws, again occurring in the weak interactions. One interpretation of the Princeton results is that there is a small (one-part-in-a-thousand) violation of time-reversal conservation in the weak interactions. Other experiments are currently in progress at laboratories around the world to find out if any further violations of the symmetry principles or conservation laws exist.

This then is the present and rather confusing state of the forefront field of research in physics—particle physics. Tremendous progress has been made in organizing the many elementary particles into groups by Gell-Mann's "Eight-Fold Way," but we do not seem to be even close to finding out why particular particles occur this way or how they interact with one another in the strong interactions. And what does the violation of parity and the partial violation of time reversal in the weak interactions have to do with all of this? At present no one knows, but most physicists hope, indeed expect, that these pieces of the puzzle will begin to fit together in the near future.

Questions

1 For what reasons do you think the term "elementary particles" was coined?

2 Do you feel that the discovery of antiparticles in nature "proves" Dirac's theory of the electron? Does the fact that it also predicts electron spin "prove" the theory?

3 Are the "laws of particle physics" natural laws or do they actually pertain to man's *description of* nature? Is there a difference between natural laws and man's interpretation of nature?

4 Dirac's equation leads to the description of phenomena with mathematical objects other than ordinary numbers. Do you think describing nature with unusual mathematical structures is valid or invalid? Explain your answer.

5 According to the picture developed in quantum electrodynamics, charged particles interact by exchanging photons. How does this picture differ from the description of the interaction of charged particles in an electric field?

6 Name some of the new quantum numbers invented to explain the phenomena of particle physics. What is the significance of these quantum numbers?

7 What do physicists mean by "symmetry"? Can you relate the concept of symmetry in physics to other more ordinary ideas of symmetry in nature?

8 How are quantum numbers and conservation laws related to the concept of symmetry?

9 Explain how Gell-Mann's Eight-Fold Way helps to classify the "fundamental" particles.

10 Why was the discovery of the Ω^- particle particularly important? Wasn't it simply one more addition to a long and constantly growing list of elementary particles?

11 What is a quark? Why were quarks "invented"?

12 If quarks are actually found to exist, will it be correct to call them "elementary" or fundamental particles? Why or why not?

Questions Requiring Outside Reading

1 Do you think that the problems in the field of particle physics will lead to a revolutionary new picture of the universe? Explain.

2 If experimentalists had failed to discover the antiparticle, would Dirac's electron theory have been destroyed? Can you think of a theory which has led to false physical predictions?

3 When Dirac discovered his theory of the electron he was 26. When Gell-Mann proposed the concept of "strangeness" he was 25. Many other great physicists have completed their most important work at a very young age. (As we have seen, most of Newton's work was done when he was 23–24 years old.) How old were some of the other physicists mentioned in this book when they did their important work? Discuss the advantages a younger man may have in approaching problems.

4 What do physicists mean when they speak of "time reversal"? Are they talking about an actual physical process?

5 Compare the proliferation of "elementary" particles in recent years with the snow-balling complexity of the Ptolemaic cosmological picture that was swept away by the revolutionary ideas of Copernicus, Galileo, and Newton.

Problems

1 You are given quarks a, b, and c. If

quark a has mass 1 unit and charge $+\frac{2}{3}$,
quark b has mass 1 unit and charge $-\frac{1}{3}$,
and
quark c has mass 2 units and charge $-\frac{1}{3}$,

how many "elementary" particles can you form from these three quarks? What particles will they be? (*Example:* (aaa), ...)

2 Construct a mass-charge diagram for all of the particles formed from the quarks in Problem 1. What is the strangeness of each "family"?

Fill in the rest of the table.

Example:

		Charge				Strangeness
		−1	0	1	2	
	6	(ccc)				
Mass	5					
	4					
	3					

3 The following reaction is observed:

$$\pi^- + p^+ \rightarrow K^0 + \Lambda^0$$

The strangeness of both the π^- and the p^+ is 0. If the strangeness of the K^0 is $+1$, what is the strangeness of the Λ^0 particle?

4 The Λ^0 particle in Problem 3 can decay as follows:

$$\Lambda^0 \rightarrow p^+ + \pi^-$$

What conservation law is violated by this decay?

5 The Ω^- has a strangeness of -3. Write the possible reactions by which the Ω^- may be produced from K^--mesons bombarding protons.

6 Write the possible reactions by which the Ω^- might decay into π-mesons and a neutron or proton.

21

Epilogue

21-1

Introduction

In this final chapter we will speculate on the future of physics—particularly the future of particle physics. To some degree, historical perspective can help to direct us. Certainly history will not repeat itself, but there may be patterns in previous revolutions that we can also recognize in the development of elementary particle physics. It is clearly impossible for us to give any definite answers to the unfinished problems and questions raised in Chapter 20; we can only select the current problems we feel are relevant and reveal the difficulties that we feel may lie in solving them.

21-2

Relativity, Quantum Mechanics, and the Elementary Particle Concept

Both the earlier Newtonian and Maxwellian revolutions arose from the synthesis of two apparently completely different fields of study. Newton synthesized terrestrial and celestial motion; Maxwell synthesized light (optics) and electromagnetism. Today another major revolution seems near which may have to do at least in part with a synthesis of quantum mechanics and relativity. Dirac's attempt at such a synthesis in 1928 resulted in remarkable success in the understanding of electrons, positrons, and photons, but these were only three of the several hundred "elementary" particles. There is a need for an even deeper understanding of the relationship between relativity and quantum mechanics. On one hand, there appears to be a crisis—a basic incompatibility between the two fields—involving the concept of "elementarity" for particles. On the other hand, there are certain similarities in the methods each field employs to deal with the particle observation process.

The crisis in the elementary particle concept arises in the following way. Quantum mechanics tells us that the size or wavelength of a particle is given by de Broglie's relation $\lambda = h/p$, where p is the momentum. However, relativity tells us that the energy of a particle at rest is $E = mc^2$ and that its rest "momentum" is E/c or $p = mc$. Thus the "size" of a particle is h/mc. Now the electrons of which a large object like an atom is composed have a size $h/m_e c$, where m_e is the electron mass. This size is smaller by a factor of about 100 to 1000 than the atom. Likewise, in a nucleus, the size of the nucleon (the neutron and the proton) is still about 10 times smaller than the nucleus. Thus we can speak of an atom as "composed of" electrons and a nucleus as "composed of" nucleons (that is, they are composed of more elementary or smaller building blocks).

However, graduating to smaller masses, we eventually reach a point where the constituent elementary particles have a larger wavelength (size) h/mc than the object they are supposed to form. So the dilemma quantum mechanics and relativity present us with is finding the actual definition of an "elementary particle." Since the size of a particle becomes larger as its mass becomes smaller, it appears impossible to "construct" a particle out of more elementary and less massive particles.

For instance, consider the quark theory of the baryons. The proton in this theory is allegedly composed of three quarks, but the size of the proton as measured experimentally is less than 10^{-13} cm (which is about the same as its characteristic wavelength $h/m_p c$). If the three quarks are to be the building blocks of the proton, each quark would presumably have to be smaller than the size of the proton. But if the size of each quark $h/m_q c$ is smaller than the proton $h/m_p c$, then it is clear that the mass of each quark is larger than the mass of the proton. So we end up with a particle (the proton) composed of three particles (quarks) whose total mass is at least three times, and probably much more than three times, the mass of the original proton. Perhaps the binding energy of three quarks is tremendously large so that three "heavy" (and therefore smaller) quarks can make up a proton. But then the binding energy between the three quarks would have to be very much larger than the total rest mass energy of the resulting proton. And how could the proton be "made up" of three particles when most of the mass energy of the three quarks, and hence their identity as three particles, is lost as binding energy. Further, the proton should be able to be separated into three quarks by high-energy collisions in a large proton accelerator, but, to date, no quarks have been found. Our concepts of "particle" and of "constituents of a particle" given by relativity and quantum mechanics appear to be in trouble.

Conversely, some similarity between relativity and quantum mechanics does exist, since each stresses new roles for the observer of physical phenomena. Relativity shows us that the state of motion of the observer makes lengths, times, and masses change by the factor $\sqrt{1 - v^2/c^2}$. (Due to this relative motion alone, we see an electric phenomenon which we now interpret as magnetism, but which, in reality, has simply been Coulomb's law all along.) So the *state of motion* of the observer is important in affecting what is measured in the "outside world." The uncertainty principle in quantum mechanics tells us that, in the very act of observing electrons by bombarding them with, say, photons, we disturb the measurement of the momentum of the electron. In other words, the *observer himself,* in the act of observing, is also affecting what is measured in the "outside world."

Somehow our act of observation, previously considered passive, may control the reality that we measure. And what is the external world—the "outside world"? Certainly, as Descartes showed so clearly 300 years ago, all we know that exists for certain is our own conscious thought and experience. Our experiences—our acts of observations—and our consciousness of those acts somehow influence what is generally regarded as "real." According to Einstein:

> *Everyone has experienced that he has been in doubt whether he has actually experienced something with his senses or has simply dreamed about it. Probably the ability to discriminate between these alternatives first comes about as the result of an activity of the mind creating order.*
>
> *Let us consider an example. A person A ("I") has the experience 'It is lightning.' At the same time the person A also experiences such a behavior of the person B as brings the behavior of B into relation with his [A's] own experience 'It is lightning.' Thus it comes about that A associates with B the experience 'It is lightning.' For the person A the idea arises that other persons also participate in the experience 'It is lightning.' 'It is lightning' is now no longer interpreted as an exclusively personal experience, but as an experience of other persons (or eventually only as a 'potential experience'). In this way arises the interpretation that 'It is lightning,' which originally entered into the consciousness as an 'experience,' is now also interpreted as an (objective) 'event.' It is just the sum total of all events that we mean when we speak of the 'real external world.'* [1]

But some physicists (and psychologists) would go even further than this. We know from the uncertainty principle that the very act of observing, of knowing or being conscious of an atomic event, affects the outcome of that event. The wave function of

---

[1] Albert Einstein, *Ideas and Opinions* (New York: Crown Publishers, Inc.). Copyright © 1954, p. 363. Used by permission of Crown Publishers, Inc.

the process changes due to the observation. Eugene Wigner, a pioneer in nuclear physics and the recipient of the Nobel Prize in 1963, stated

> . . . the impression which one gains at an interaction may, and in general does, modify the probabilities with which one gains the various possible impressions at later interactions. In other words, the impression which one gains at an interaction, called also the result of an observation, modifies the wave function of the system. The modified wave function is, furthermore, in general unpredictable before the impression gained at the interaction has entered our consciousness: It is the entering of an impression into our consciousness which alters the wave function because it modifies our appraisal of the probabilities for different impressions which we expect to receive in the future. It is at this point that the consciousness enters the theory unavoidably and unalterably.[2]

Must we then have to find a theory of physics that takes human consciousness into account? Perhaps. But certainly our ideas on observation and cause and effect (which depends upon the separation of the observed from the observer) may have to undergo radical changes before we can completely understand what particle physics is all about. A problem can already be seen in our ideas of one clump of matter (the observer) observing another clump of matter (the observed object). Geoffrey Chew, a theoretical physicist at Berkeley, has argued that as long as we can speak of atoms as "composed" of electrons and of nuclei as "composed" of neutrons and protons, we can still talk of one clump of matter "observing" other clumps of matter. Our experimental apparatus observes the experiment. But when this breaks down, as it eventually must do when the size becomes very small, we may have to abandon the usual concepts of observation of matter. Perhaps we can no longer separate or compartmentalize the observer and the observed. We may have to solve the *entire* problem, taking both the observer and the observed quantity—and perhaps the entire universe as well—into account.

21-3

Other Problems

Language is one of the many difficulties still facing those involved in the new revolution in physics. When we do not know what the new "rules of the game" [or paradigms, as Kuhn (see below) calls them] are, we may not even have the necessary terms at our disposal with which to express the new conceptual ideas. We must see and explain the familiar data in a new way, in the context of a new framework, but the resistance to finding and accepting

[2] I. J. Good, A. J. Mayne, and J. M. Smith, eds., *The Scientist Speculates* (New York: Basic Books, Inc., Publishers, 1962), p. 98.

a new paradigm is enormous. Even the experiments we choose to conduct are related to our own world view—to the old paradigm. T. S. Kuhn, in *The Structure of Scientific Revolutions*, writes:

> Science does not deal in all possible laboratory manipulations. Instead, it selects those relevant to the juxtaposition of a paradigm with the immediate experience that that paradigm has partially determined.[3]

Language can trap us in the conceptual framework of the old paradigm so that it is extremely difficult, if not impossible, to "see" the data in a new way. A phrase like "the sun sets" attributes motion to the sun—the Ptolemaic conceptual picture. Likewise, "particle and wave" relate to two separate classical conceptual experiences, but an actual electron is both particle and wave. Heisenberg states this in an unequivocal way:

> The solution of the difficulty is that the two mental pictures which experiments lead us to form—the one of particles, the other of waves—are both incomplete and have . . . the validity of analogies which are accurate only in limiting cases. It is a trite saying that 'analogies cannot be pushed too far,' yet they may be justifiably used to describe things for which our language has no words. Light and matter are both single entities, and the apparent duality arises in the limitations of our language.
>
> It is not surprising that our language should be incapable of describing the processes occurring within the atoms, for, as has been remarked, it was invented to describe the experiences of daily life, and these consist only of processes involving exceedingly large numbers of atoms. Furthermore, it is very difficult to modify our language so that it will be able to describe these atomic processes, for words can only describe things of which we can form mental pictures, and this ability, too, is a result of daily experience.[4]

The parable of the cave in Plato's *Republic* gives another example of the difficulties involved in finding and expressing a new world view. The men chained in Plato's cave can only see the shadows of themselves which the fire behind them throws on the opposite wall of the cave. They cannot see each other or their own bodies. This is their entire world—a world of two-dimensional shadows. But the men have been there all their lives and have always thought and talked of these shadows as if they were real. They have no ideas or language with which to express a concept of themselves—the objects that produce the shadows. A man who escapes out into the sun and sees real three-dimensional

---

[3] T. S. Kuhn, *The Structure of Scientific Revolutions* (Chicago: University of Chicago Press, 1970), p. 126. Used by permission.

[4] Werner Heisenberg, *The Physical Principles of the Quantum Theory* (Chicago: University of Chicago Press, 1930), p. 13. Used by permission.

things for the first time will have an exceedingly difficult time convincing those in the cave that their "reality" of shadows is false. He won't even have a language with which to describe his new three-dimensional world. His task would be so difficult that he might not even choose to return to the cave at all.

The problem of language in revolutions in physics has been discussed by one of the philosophers of modern physics, the late N. R. Hanson:

> When language and notation are ignored in studies of observation, physics is represented as resting on sensation and low-grade experiment. It is described as [a] repetitious, monotonous concatenation of spectacular sensations and of school-laboratory experiments. But physical science is not just a systematic exposure of the senses to the world; it is also a way of thinking about the world, a way of forming conceptions. The paradigm observer is not the man who sees and reports what all normal observers see and report, but the man who sees in familiar objects what no one else has seen before.[5]

The new paradigm must not only account for all past experiments; it must explain new and as yet unanswered ones as well. This makes new paradigms difficult to find, but it also creates an excitement that comes from looking for a new world view—the excitement of working with physics.

21-4

Conclusion

There is so much that we now know about our universe—from the very big to the very small. And yet there is so much we still do not know. Why is the gravitational force and the electrical force an inverse square force? Why are there two kinds of electrical force—attraction and repulsion? Why do "particles" behave like "waves"? Why is the velocity of light measured as the same by all observers, regardless of their state of motion? Why does Gell-Mann's "Eight-Fold Way" work so well? Why are parity and time reversal symmetries violated in weak but not in strong interactions? Why should the center of charge of a family of particles determine its strangeness quantum number? Why are there three dimensions of space and only one of time?

Some scientists may consider these questions nonscientific, mystical, or metaphysical, and therefore downgrade them, but Einstein did not. He once said:

> The most beautiful and most profound emotion we can experience is the sensation of the mystical. It is the sower of all true science. He to whom this emotion is a stranger, who can no longer wonder

[5]Norwood R. Hanson, *Patterns of Discovery: An Enquiry into the Conceptual Foundations of Science* (Cambridge: Cambridge University Press, 1958–1965), p. 30. Used by permission.

and stand rapt in awe, is as good as dead. To know that what is impenetrable to us really exists, manifesting itself as the highest wisdom and the most radiant beauty which our dull faculties can comprehend only in their most primitive forms—this knowledge, this feeling is at the center of true religiousness.[6]

The beauty and awe that a scientist feels in "discovering" the workings of the universe is akin to the feelings a painter, musician, or poet experiences while creating his work. There is only one universe to discover and it can be discovered for the first time only once—reason enough for the excitement modern physicists experience as they pursue the frontiers of particle physics. For this pursuit is truly, as Richard Feynman has called it, "the grand adventure."

[6]Lincoln Barnett, *The Universe and Doctor Einstein,* p. 108. Copyright © William Morrow & Co., Inc., 1968. Used by permission.

Appendixes

In order to analyze projectile (compound) motion (when an object is neither fired straight up nor in a horizontal direction, but at some *angle with* the horizontal), it is easier if we introduce the simple concept of a vector. A vector is a quantity which has *magnitude* (length, depth, etc.) and also has a *direction in space.* We denote a vector by an arrow (the length giving its magnitude, the shaft and head its direction). Earlier we introduced velocity as both $v = s/t$ and $v = at$ without referring to any direction. This was not strictly correct.

In a moving car, for instance, we have a velocity at any given moment that has both direction (east, north, south, west, etc.) *and* magnitude (which we will call *speed* from now on). Thus velocity should be a vector. We will denote a vector quantity by $\vec{v}$ and its magnitude by v (without the arrow).

If we wish to add vector $\vec{a}$ to vector $\vec{b}$ (Figure 4A-1), the rules of this mathematical game are: First place the tail of $\vec{b}$ at the head of $\vec{a}$; the resultant vector $\vec{c}$, when drawn from the tail of $\vec{a}$ to the head of $\vec{b}$, is the sum of the two vectors $\vec{a}$ and $\vec{b}$ (see Figure 4A-2).

The difference between two vectors can be defined in a similar manner. The difference $\vec{a} - \vec{b}$ (Figure 4A-3) is obtained by adding the vector $(-\vec{b})$ to $\vec{a}$ to obtain $\vec{c}$ as before (see Figure 4A-4).

In compound motion we use the ideas behind vector summation to separate the vertical part of an initial velocity from the horizontal part. For instance, if a shell is fired into the air with an initial velocity of 500 ft/sec *at an angle of 45°* (see Figure 4A-5),

Appendix 4A

Mathematical Addendum on Vectors

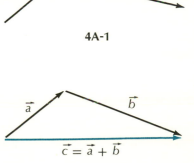

4A-1

$$\vec{c} = \vec{a} + \vec{b}$$

4A-2

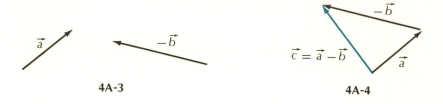

4A-3

4A-4

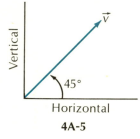

4A-5

then we can easily compute both that part of 500 ft/sec which is the upward (vertical) part of the initial velocity and that which is the horizontal part. We need two vectors, $\vec{v}_h$ for the horizontal and $\vec{v}_v$ for the vertical, which vectorially add up to the original $\vec{v}$. The solution to our problem is obvious (see Figure 4A-6).

Can we compute $\vec{v}_v$ to use in Equation (4-7), page 46, to calculate the vertically accelerated motion? Yes, since $\vec{v}_v$ and $\vec{v}_h$ always form a right angle, we can use the Pythagorean theorem ($a^2 + b^2 = c^2$) if we know the angle $\vec{v}$ makes. In the example given, the 45° triangle gives sides in the ratios: $\vec{v}:\vec{v}_h:\vec{v}_v = \sqrt{2}:1:1 = 1:0.707:0.707$. So if $\vec{v}$ has a magnitude of 500 ft/sec, then $v_h = 500 \times 0.707 = 354$ ft/sec. Likewise, $v_v = 354$ ft/sec. For angles other than 45°, we use the same principle.

Since velocity is a vector and time is not, the definition of acceleration in Equation (4-4), page 44, should yield a vector also. This in indeed true. We already know that acceleration can be in the same direction as the velocity or in the opposite direction (deceleration). Furthermore, $\vec{a}$ does not have to line up with $\vec{v}$, but can make an angle with $\vec{v}$, in which case the acceleration can change not only the magnitude of $\vec{v}$ but also its direction!

If $\vec{a}$ and $\vec{v}$ are not in the same direction, we need to modify Equation (4-4) to allow for this. If the velocity changes slightly in a small interval of time, we will call the small change in velocity $\Delta\vec{v}$. (The symbol "Δ" means "a small change in." Thus $\Delta\vec{v}$ means "a small change in $\vec{v}$.") This change in velocity occurs in a time interval which is also small and which we will call Δt. Then acceleration is defined as

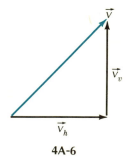

4A-6

$$\vec{a} = \frac{\Delta\vec{v}}{\Delta t}$$

if the time interval is thought of as being as small as possible. This is an *instantaneous acceleration*. Remember $\Delta\vec{v}$ is the *change* in velocity and is really the difference between the velocity *vector before* and the velocity *vector after* the time interval Δt is measured. Symbolically,

$$\vec{a} = \frac{\vec{v_2} - \vec{v_1}}{t_2 - t_1} = \frac{\Delta \vec{v}}{\Delta t}$$

where "1" and "2" refer to "before" and "after" the velocity change, respectively.

Example 4A-1:

A sports car headed due north at 30 ft/sec executes a sharp right turn and $\frac{1}{2}$ sec later is headed due east at the same speed. What was the average acceleration during the turn? Geometrically,

velocity before $= \vec{v_1}$ $\uparrow v_1$

velocity after $= \vec{v_2}$ $\underset{\rightarrow}{v_2}$

Now $\vec{a} = \dfrac{\vec{v_2} - \vec{v_1}}{t_2 - t_1} = \dfrac{\Delta \vec{v}}{\Delta t}$ [$\vec{v_2} - \vec{v_1}$ is obtained by adding $(-\vec{v_1})$ to $\vec{v_2}$]

By the Pythagorean theorem, the magnitude of the resultant $\vec{v_2} - \vec{v_1}$ is the magnitude of $\vec{v_1}$ (or $\vec{v_2}$) $\times \sqrt{2} = 30\sqrt{2} = 42.4$ ft/sec. The average acceleration is just this magnitude divided by the time interval

$$\Delta t = t_2 - t_1 = \tfrac{1}{2} \text{ sec or } \vec{a} = \frac{42.4}{\frac{1}{2}} = \mathbf{84.8 \ ft/sec^2}$$

We see that the direction of $\vec{a}$ is the same as $(\vec{v_2} - \vec{v_1})$ and is seen from Figure 4A-7 to be southeast.

(In this example, we calculated only the *average* acceleration. Actually $\vec{a}$ changes in direction all the time during the turn. If we had used smaller time intervals, this would have been apparent in the calculation.)

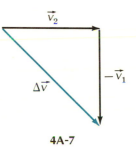

4A-7

The use of vectors is not the only unique way to work with accelerations, velocity, etc. Motion can be described without vectors. Projectile trajectories can be calculated without vectors if we keep separate records of the horizontal motion and the vertical motion. However, vectors are useful in that they combine these two components of motion in a simple way. We will use vectors more and more as we follow the major ideas in physics.

It is often useful to have a geometric or graphic picture of an algebraic formula. In Figure 4B-1, we plot velocity along the vertical axis and time along the horizontal axis for Equation (4-2) $s = vt$, the distance traversed at constant velocity. For Example 4-2 (page 42), the constant velocity of 1000 mi/hr is represented by the horizontal straight line at the value of $v = 1000$ mi/hr. The distance s increases proportionally with time t, and the quanti-

Appendix 4B

Graphic Representation of Velocity and Acceleration Formulas

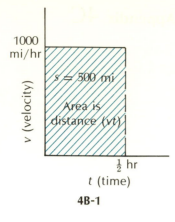

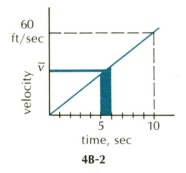

4B-1

4B-2

tative measure of s is given by the area $v \times t$, which is the area of a rectangle. For $t = \frac{1}{2}$ hr, the distance $s = 500$ mi is represented by the area of the colored portion of the figure.

If the velocity is not constant, however, as is the case in most examples of physical interest, the line depicting velocity will not be horizontal. Constant velocity in the physical world can only to be applied to the motion of objects in a horizontal plane. In this case, Equation (4-2) agrees with experiment since we observe that equal distances are traversed in equal times.

Figure 4B-2, a graph of velocity with constant acceleration, plots the result of $s = \frac{1}{2}at^2$ [Equation (4-8), page 47] in which an object starts from rest and accelerates to 60 ft/sec in 10 sec. The velocity, represented by the colored diagonal, increases by equal increments in equal times. What then is the distance traveled? The distance traveled in a given, small time interval (between 5 sec and 6 sec, for example) is just the average velocity, call it $\bar{v}$ (approximately constant over a small enough time) multiplied by the interval of 1 sec (in this case). Thus the distance traveled between 5 and 6 sec is represented by the colored area on the graph. The total distance traveled in accelerating to 60 ft/sec in 10 sec is therefore the sum of all increments like this colored region or simply the area of the triangle under the diagonal line. The area of a right triangle is $A = \frac{1}{2}bh$ (here $b = $ base $= $ time and $h = $ height $= $ velocity). So, in our case,

$$s = \frac{1}{2}vt$$

but for constant acceleration $v = at$ by definition, so

$$s = \frac{1}{2}at^2$$

which agrees with Equation (4-8). If we now have an initial velocity v_0 which is not zero, the graph must start at zero time with a velocity of v_0. A short analysis of the two graphs in Figure 4B-3 shows that total area $= s = v_0 t + \frac{1}{2}at^2$ which agrees exactly with the algebraic analysis leading to Equation (4-7), page 46.

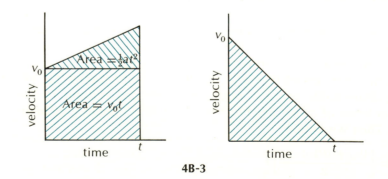

4B-3

Unit	mks System	cgs System	English System
length (s)	meter (m) 1 m	centimeter (cm) = 100 cm	foot (ft) = 3.1 ft
time (t)	second (sec)	second	second
velocity (v)	m/sec 1 m/sec	cm/sec = 100 cm/sec	ft/sec = 3.1 ft/sec
acceleration (a)	m/sec² 1 m/sec²	cm/sec² = 100 cm/sec²	ft/sec² = 3.1 ft/sec²

Under constant acceleration, velocity, by definition, is always increasing linearly from v_0 to v. As a result, the average velocity $\bar{v}$ will be half of the sum of the initial and final velocities:

$$\bar{v} = \frac{v_0 + v}{2}$$

But by definition the average velocity must also be the *total* distance s divided by the total time t:

$$\bar{v} = \frac{s}{t}$$

We set these two expressions for $\bar{v}$ equal to one another and we obtain

$$\frac{s}{t} = \frac{v_0 + v}{2}$$

Or, multiplying by t on each side, we obtain the distance s in the case of *constant acceleration only*:

$$s = \left[\frac{v_0 + v}{2} \right] t$$

[Note the difference between this equation for constant acceleration and Equation (4-2) for constant velocity.]

This relation is between the four quantities s, v_0, v, and t. Notice that acceleration a is not present, even though the above equation holds only for constant *accelerated* motion. To obtain s in terms of a, we use Equation (4-5) to eliminate v in favor of a. Then we obtain,

$$s = \left[\frac{(v_0 + at) + v_0}{2} \right] t$$

and, by multiplying the terms within the brackets by t and collecting terms, we obtain the desired relation for s in terms of v_0, a, and t or Equation (4-7):

$$s = v_0 t + \tfrac{1}{2}at^2$$

Appendix 5

Derivation of Centripetal Force, Equation (5-1)

The dashed line in Figure 5A-1 shows a part of the circular orbit of the moon. Let the moon-earth distances AE and CE each be called r. After a period of time t, the moon traveling at constant speed v would be expected to reach B, thus covering a distance vt from A to B. But the moon is constantly "falling" toward the earth due to centripetal acceleration a_c. Therefore, in the same time t it took to go from A to B, the moon will also have fallen the distance BC.

If the time t is very small, the moon "falls" only a very small distance and we can assume that, over such small distances, the acceleration of gravity is constant. But for constant acceleration, we know from Equation (4-8) (page 47) that $BC = \tfrac{1}{2}a_c t^2$.

ABE in Figure 5A-1 forms a right triangle (shown in greater detail in Figure 5A-2). Using the Pythagorean theorem $a^2 + b^2 = c^2$, where the hypotenuse BE is $(r + \tfrac{1}{2}a_c t^2)$, we have a relation we can solve for a_c:

$$(r + \tfrac{1}{2}a_c t^2)^2 = (vt)^2 + r^2$$

Squaring the terms in parentheses gives us:

$$r^2 + \tfrac{1}{4}a_c^2 t^4 + a_c r t^2 = v^2 t^2 + r^2$$

r^2 cancels out on each side and we can cancel a factor of t^2 in each remaining term to obtain:

$$\tfrac{1}{4}a_c^2 t^2 + a_c r = v^2$$

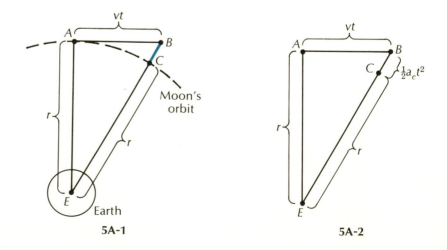

5A-1

5A-2

Now, if we make t vanishingly small, the term $\frac{1}{4}a_c^2 t^2$ becomes zero in comparison with the other terms, and we have

$$a_c r = v^2$$

or

$$a_c = \frac{v^2}{r} \qquad (5\text{-}1)$$

for the value of the centripetal acceleration required to keep an object in a circular path.

We set the centripetal force in Equation (6-7) equal to the gravitational force in Equation (6-8) (pages 80–81):

$$F_c = F_{\text{grav}} = G\frac{Mm}{R^2} = \frac{mv^2}{R}$$

We now cancel the planet's mass m from both sides of the equation and express the orbital velocity v in terms of the time period T. Since $v = $ total distance/total time,

$$v = \frac{2\pi R}{T}$$

which, when substituted into the first equation above, yields

$$G\frac{M}{R^2} = \frac{(2\pi R)^2/T^2}{R}$$

If we multiply both sides of this equation by R^2 and divide both sides by $(2\pi)^2$, we find the result of Equation (6-9) (page 81):

$$\frac{GM}{(2\pi)^2} = \frac{R^4}{RT} = \frac{R^3}{T^2}$$

Appendix 6

Derivation of Kepler's Third Law

We begin with the change in KE as the product of the force F with the distance s

$$KE = F \times s$$

and use $s = \frac{1}{2}at^2$ for constant force or acceleration:

$$KE = F \times \frac{1}{2}at^2$$

Using Newton's second law $F = ma$, we have

$$KE = \frac{1}{2}ma^2 t^2$$

which becomes, by substituting $v = at$,

$$KE = \frac{1}{2}mv^2$$

Appendix 7

Derivation of Kinetic Energy, Equation (7-2)

Appendix 14

Proof that the Lorentz Transformations Satisfy the Relativity Equation

To prove that Equations (14-6)–(14-9) satisfy Equation (14-5), we substitute x' in Equation (14-8) and t' in Equation (14-9) into Equation (14-5) in the same way in which we unsuccessfully tested the Galilean transformation Equations (14-1) and (14-2) earlier. Then Equation (14-5)

$$x^2 - c^2 t^2 = x'^2 - c^2 t'^2 \qquad\qquad \text{(14-5)}$$

becomes

$$x^2 - c^2 t^2 = \left(\frac{x - vt}{\sqrt{1 - v^2/c^2}}\right)^2 - c^2 \left(\frac{t - vx/c^2}{\sqrt{1 - v^2/c^2}}\right)^2$$

Note that the common denominator $(\sqrt{1 - v^2/c^2})^2$ on the right-hand side of the equation becomes $1 - v^2/c^2$. Then

$$x^2 - c^2 t^2 = \frac{(x - vt)^2 - c^2[(t - vx)/c^2]^2}{1 - v^2/c^2}$$

Next, we expand each squared term within parentheses:

$$x^2 - c^2 t^2 = \frac{x^2 - 2vxt + v^2 t^2 - c^2(t^2 - 2vxt/c^2 + v^2 x^2/c^4)}{1 - v^2/c^2}$$

and multiply c^2 into each term inside the parentheses:

$$x^2 - c^2 t^2 = \frac{x^2 - 2vxt + v^2 t^2 - c^2 t^2 + 2vxt - v^2 x^2/c^2}{1 - v^2/c^2}$$

Note that $-2vxt$ cancels $+2vxt$. Next collect the two terms on the right-hand side which contain x^2 and the two terms which contain t^2:

$$x^2 - c^2 t^2 = \frac{x^2 - x^2 v^2/c^2 - c^2 t^2 + v^2 t^2}{1 - v^2/c^2}$$

We can factor

$$x^2 - x^2 v^2/c^2 = x^2(1 - v^2/c^2)$$

and

$$-c^2 t^2 + v^2 t^2 = -c^2 t^2 + (v^2/c^2)c^2 t^2 = -c^2 t^2(1 - v^2/c^2)$$

Then we are left with

$$x^2 - c^2 t^2 = \frac{x^2(1 - v^2/c^2) - c^2 t^2(1 - v^2/c^2)}{1 - v^2/c^2}$$

Finally, we factor

$$x^2(1 - v^2/c^2) - c^2 t^2(1 - v^2/c^2) = (x^2 - c^2 t^2)(1 - v^2/c^2)$$

and find

$$x^2 - c^2 t^2 = \frac{(x^2 - c^2 t^2)(1 - v^2/c^2)}{1 - v^2/c^2}$$

or

$$x^2 - c^2t^2 = x^2 - c^2t^2$$

Therefore, the new transformations of Equations (14-6) and (14-7) or of Equations (14-8) and (14-9) do indeed satisfy Equation (14-5) and must therefore be the correct solution of Equation (14-5). In other words, these new transformations of distances and times from a stationary observer A to a moving observer A' or vice versa yield the result of Equation (14-5): The velocity of light is independent of the velocity of the observer of the light source.

The velocity u of an object as measured by A' in a moving reference system is simply

$$u = \frac{x'}{t'}$$

where x' and t' are the distance and time as measured at rest with A'. Now if A' is moving past observer A with a velocity v, the velocity of the same object according to A is

$$\text{velocity sum} = W = \frac{x}{t}$$

where the length x corresponds to x' and the time interval t corresponds to t'. Using Equation (14-6) for x in terms of x' and t', and Equation (14-7) for t (page 186), we find, by substituting into the preceding equation, that

$$\text{velocity sum} = W = \frac{\dfrac{x' + vt'}{\sqrt{1 - v^2/c^2}}}{\dfrac{t' + vx'/c^2}{\sqrt{1 - v^2/c^2}}}$$

Note that the factor $\sqrt{1 - v^2/c^2}$ cancels in the numerator and denominator, so that

$$W = \frac{x' + vt'}{t' + vx'/c^2}$$

Now we divide both terms in the numerator and both terms in the denominator by t' and find

$$W = \frac{x'/t' + v(t'/t')}{t'/t' + (vx'/t')/c^2}$$

But since $x'/t' = u$ and $t'/t' = 1$, we have

$$W = \frac{u + v}{1 + (vu/c^2)}$$

which is the velocity addition formula.

Appendix 15

Derivation of the Velocity Addition Formula (Section 15-4)

Appendix 16

Derivation of Equation (16-3)

If we square Equation (15-4) (page 198)

$$m'^2 = \frac{m^2}{1 - v^2/c^2}$$

and then multiply both sides of the equation by $(1 - v^2/c^2)$, we have

$$m'^2 - m'^2 \frac{v^2}{c^2} = m^2$$

Finally, we multiply both sides of the equation by c^4 to obtain

$$m'^2c^4 - m'^2v^2c^2 = m^2c^4$$

or

$$(m'c^2)^2 - (m'v)^2c^2 = m^2c^4$$

We note that $m'c^2 = E$ and $m'v = p$, giving us the energy-momentum equation

$$E^2 - p^2c^2 = m^2c^4$$

Index

Boldface numbers refer to biographical sketches.

Inclined plane, 46–48, 87
Induction, electromagnetic, 136–41, 160, 162
Inertia, 29, 52–53, 74, 78–80, 209
Interference:
  of electrons, 227
  of light, 150–51, 172–73
  of waves, 150
Interferometer, Michelson's, 168
Interval in space-time, 205
Isotope, 260

Jensen, Hans, 264

Kaon (see K-meson)
Kelvin, Lord, 166–67, 214, 268
Kepler, Johannes, 23, 33, 55, 59–68, **65,** 71, 88, 174, 219, 281
Kepler's laws, 63–64, 80–82
Kepler's solar system, 66
Kinetic energy, 96, 98–100, 200–201, 216
K-meson, 277–84, 286
Kuhn, Thomas, 294

Lambda particle, 277–84
Language and physics revolutions, 268, 293–95
Lee, T.D., 286
Length contraction in relativity, 193–95, 203, 207
Length, proper, 195
Lepton, 276
Levity, 13
Light, 144–54
  bending of, 210
  constancy of speed of, 187–88
  dispersion of, 144
  interference of, 150–51, 172–73
  particle theory (photons), 144–45, 215–16, 231
  polarization of, 152–53
  transverse waves, 151
  velocity of, 153, 167, 201
  wave theory, 145, 146–49, 153

Liquid drop model, 260, 264
Lithium, 248
Longitudinal wave, 152
Lorentz, H. A., 186
Lorentz transformations, 186, 204–205
Luther, Martin, 24

Mach, Ernst, 92
Magnet-current forces, 133
magnetic fields, 132–36, 153, 156, 207
  and electric current forces, 132–36
Magnetic poles, 160
Magnetism, 104, 132–36, 292
  and currents, 121
  and Oersted, 121–24
Magnet-magnet forces, 133
Mass, 74
  energy, 80, 100, 199–201
  gravitation, 79–80
  inertia, 78–80
  relativistic increase in, 198, 203
  rest, 198, 200–201
  weight, 84
Matrix, 271
Maxwell, James Clerk, 154, 156–64, **158,** 167, 173–74, 268, 290
  implications of, in electromagnetic revolution, 164, 173–74
  and Newton, 159
  synthesis of, 157
Maxwell's equations, 159–63, 166, 174, 178–80, 182
  with magnetic field equations, 159–62
  without magnetic field equations, 162
Mayer, Maria, 264
Mechanical energy, 96
Mechanical Revolution, 23–92
Mechanistic models of universe, 1–14
Meitner, Lise, 260
Mendeleev, Dimitri, 220–21, 248
Meson, 276–84

Michelson, Albert C., 164, **169,** 167–74
  interferometer, 168
Michelson-Morley experiment, 167–74, 176, 184, 214
  results of, 173
mks system (*see* Units)
Models:
  of atom, 222–29
  of nucleus, 263–65
  of universe, 1–13, 16–21, 23–34
Momentum, 93
  angular, 100, 224–25, 250
  conservation of angular, 101
  conservation of linear, 93–98
  quantization of angular, 224–25, 250
  and relativity, 196–99
Motion:
  accelerated (*see* Acceleration)
  Aristotle's laws of, 13–15
  centripetal, 67–68
  changes in (acceleration), 44
  compound, 50–52
  decelerated, 38, 45
  diurnal, 24
  of earth through the ether, 167–68, 173
  forced, 13
  Galileo's concepts of, 36–38
  natural, 13, 37
  periodic, 147–48
  projectile, 14, 50
  quantitative descriptions of, 42–43
  relative, 30
  resistance to, 37–38
  retrograde, 18, 20, 24, 26, 34
  speed (*see* Velocity)
  velocity, 41
Muon, 274
Mysticism, 1, 5, 65, 295–96
Mythology, 1

Natural motion, 13–14, 37
Ne'eman, Yuval, 281
Neutrino, 274

Neutron, 257
Newtonian mechanics:
  and angular momentum, 100–101
  and conservation of energy, 96–98
  and conservation of momentum, 93–96
  and potential energy, 98–100
  and quantum mechanics, 214, 219, 243, 252
  and relativity, 196, 204
Newton, Isaac, 28, 52, 67–68, 71–88, **73,** 92–93, 105–106, 113, 117, 129, 144–45, 154, 157–59, 164, 173–74, 177, 268, 290
  and concept of force and mass, 74–77
  contributions to physics, 71–72
  and Maxwell, 159
Newton's laws, 77–78, 92–93, 104, 130, 166, 178–79, 196
Nuclear:
  chain-reacting systems, 262
  composition, 257
  energy, 262
  energy levels, 259
  fission, 258, 259, 260–62
  force, 258–59, 263
  fusion, 258, 262
  matter, 257
  models, 263–65
  size, 257
Nucleons, 238, 258
Nucleus of atom, 222, 257–59
Numerology, 6, 223

Observer, effects of, 290–93
  in quantum mechanics, 243–44, 252
  in relativity, 206–208
Octet of elementary particles, 281
Oersted, Hans C., 121–22, **123,** 124, 136
Optics (*see* Light)
Osiander, Andreas, 32–33

312 Index

A 3
B 4
C 5
D 6
E 7
F 8
G 9
H 0
I 1
J